MW01167494

INTEGRATING MULTISCALE OBSERVATIONS OF U.S. WATERS

Committee on Integrated Observations for Hydrologic and Related Sciences

Water Science and Technology Board

Division on Earth and Life Studies

NATIONAL RESEARCH COUNCIL
OF THE NATIONAL ACADEMIES

THE NATIONAL ACADEMIES PRESS
Washington, D.C.
www.nap.edu

THE NATIONAL ACADEMIES PRESS 500 Fifth Street, N.W. Washington, DC 20001

NOTICE: The project that is the subject of this report was approved by the Governing Board of the National Research Council, whose members are drawn from the councils of the National Academy of Sciences, the National Academy of Engineering, and the Institute of Medicine. The members of the committee responsible for the report were chosen for their special competences and with regard for appropriate balance.

Support for this project was provided by National Aeronautics and Space Administration Grant Number NNG05GK08G, National Oceanic and Atmospheric Administration, Contract Number DG133R04CQ0009, National Science Foundation Grant Number EAR-0340018, Nuclear Regulatory Commission Grant Number NRC-04-05-087, U.S. Army Corps of Engineers Contract Number W912EK-05-P-0209, and U.S. Environmental Protection Agency Cooperative Agreement Number X3-83146501. Any opinions, findings, conclusions, or recommendations expressed in this publication are those of the author(s) and do not necessarily reflect the views of the organizations or agencies that provided support for the project.

International Standard Book Number 13: 978-0-309-11457-8
International Standard Book Number 10: 0-309-11457-8

Cover: Trout Lake area photo courtesy of Carl Bowser, Silver Pixel Images, Madison, Wisconsin.

Additional copies of this report are available from the National Academies Press, 500 Fifth Street, N.W., Lockbox 285, Washington, DC 20055; (800) 624-6242 or (202) 334-3313 (in the Washington metropolitan area); Internet, http://www.nap.edu.

Copyright 2008 by the National Academy of Sciences. All rights reserved.

Printed in the United States of America.

THE NATIONAL ACADEMIES

Advisers to the Nation on Science, Engineering, and Medicine

The **National Academy of Sciences** is a private, nonprofit, self-perpetuating society of distinguished scholars engaged in scientific and engineering research, dedicated to the furtherance of science and technology and to their use for the general welfare. Upon the authority of the charter granted to it by the Congress in 1863, the Academy has a mandate that requires it to advise the federal government on scientific and technical matters. Dr. Ralph J. Cicerone is president of the National Academy of Sciences.

The **National Academy of Engineering** was established in 1964, under the charter of the National Academy of Sciences, as a parallel organization of outstanding engineers. It is autonomous in its administration and in the selection of its members, sharing with the National Academy of Sciences the responsibility for advising the federal government. The National Academy of Engineering also sponsors engineering programs aimed at meeting national needs, encourages education and research, and recognizes the superior achievements of engineers. Dr. Charles M. Vest is president of the National Academy of Engineering.

The **Institute of Medicine** was established in 1970 by the National Academy of Sciences to secure the services of eminent members of appropriate professions in the examination of policy matters pertaining to the health of the public. The Institute acts under the responsibility given to the National Academy of Sciences by its congressional charter to be an adviser to the federal government and, upon its own initiative, to identify issues of medical care, research, and education. Dr. Harvey V. Fineberg is president of the Institute of Medicine.

The **National Research Council** was organized by the National Academy of Sciences in 1916 to associate the broad community of science and technology with the Academy's purposes of furthering knowledge and advising the federal government. Functioning in accordance with general policies determined by the Academy, the Council has become the principal operating agency of both the National Academy of Sciences and the National Academy of Engineering in providing services to the government, the public, and the scientific and engineering communities. The Council is administered jointly by both Academies and the Institute of Medicine. Dr. Ralph J. Cicerone and Dr. Charles M. Vest are chair and vice chair, respectively, of the National Research Council.

www.national-academies.org

COMMITTEE ON INTEGRATED OBSERVATIONS FOR HYDROLOGIC AND RELATED SCIENCES

KENNETH W. POTTER, *Chair*, University of Wisconsin, Madison
ERIC F. WOOD, *Vice Chair*, Princeton University, New Jersey
ROGER C. BALES, University of California, Merced
LAWRENCE E. BAND, University of North Carolina, Chapel Hill
ELFATIH A.B. ELTAHIR, Massachusetts Institute of Technology, Cambridge
ANTHONY W. ENGLAND, University of Michigan, Ann Arbor
JAMES S. FAMIGLIETTI, University of California, Irvine
KONSTANTINE P. GEORGAKAKOS, Hydrologic Research Center, San Diego, California
DINA L. LOPEZ, Ohio University, Athens
DANIEL P. LOUCKS, Cornell University, Ithaca, New York
PATRICIA A. MAURICE, University of Notre Dame, West Lafayette, Indiana
LEAL A. K. MERTES, University of California, Santa Barbara (until September 2005)
WILLIAM K. MICHENER, University of New Mexico, Albuquerque
BRIDGET R. SCANLON, University of Texas, Austin

National Research Council Staff

WILLIAM S. LOGAN, Study Director
ANITA A. HALL, Senior Program Associate

WATER SCIENCE AND TECHNOLOGY BOARD

CLAIRE WELTY, *Chair*, University of Maryland, Baltimore County
JOAN G. EHRENFELD, Rutgers, The State University of New Jersey, New Brunswick, New Jersey
SIMON GONZALEZ, National Autonomous University of Mexico, Mexico City
CHARLES N. HAAS, Drexel University, Philadelphia, Pennsylvania
JAMES M. HUGHES, Emory University, Atlanta, Georgia
THEODORE L. HULLAR, Cornell University, Ithaca, New York
KIMBERLY L. JONES, Howard University, Washington, DC
G. TRACY MEHAN, The Cadmus Group, Inc., Arlington, Virginia
JAMES K. MITCHELL, Virginia Polytechnic Institute and State University, Blacksburg
DAVID H. MOREAU, University of North Carolina, Chapel Hill
LEONARD SHABMAN, Resources for the Future, Washington, DC
DONALD I. SIEGEL, Syracuse University, Syracuse, New York
SOROOSH SOROOSHIAN, University of California, Irvine
HAME M. WATT, Independent Consultant, Washington, DC
JAMES L. WESCOAT, JR., University of Illinois at Urbana-Champaign
GARRET P. WESTERHOFF, Malcolm Pirnie, Inc., Fair Lawn, New Jersey

Staff

STEPHEN D. PARKER, Director
LAUREN E. ALEXANDER, Senior Staff Officer
LAURA J. EHLERS, Senior Staff Officer
JEFFREY W. JACOBS, Senior Staff Officer
STEPHANIE E. JOHNSON, Senior Staff Officer
WILLIAM S. LOGAN, Senior Staff Officer
M. JEANNE AQUILINO, Financial and Administrative Associate
ANITA A. HALL, Senior Program Associate
ELLEN A. DE GUZMAN, Research Associate
DOROTHY K. WEIR, Senior Program Associate
MICHAEL J. STOEVER, Senior Project Assistant

Preface

This report is a product of the Committee on Integrated Observations for Hydrologic and Related Sciences. The committee was organized under the auspices of the Water Science and Technology Board (WSTB) of the National Research Council (NRC) and the WSTB's Committee on Hydrologic Science (COHS).

This study has interesting origins. The initial idea for the study arose from discussions at the COHS in 2003, and took the form of a one-page prospectus titled *Hydrology from Space*. While the prospectus mentioned integrating space-based observations with in-situ observations, as envisioned it emphasized a top-down approach. It would have assessed the usefulness of remotely sensed observations for flood and drought prediction and for snow pack measurement, and evaluated the scientific and technical readiness for observations of selected hydrologic states, among other tasks. An assurance of partial funding for the project was made at that time by the Terrestrial Hydrology Program of the National Aeronautics and Space Administration (NASA).

However, a near simultaneous effort was underway for a broader NRC "decadal survey" to help set an agenda for observations in support of *Earth Science and Applications from Space.* Among that committee's key tasks were to develop a consensus on the top-level scientific questions that should provide the focus for Earth and environmental observations for a 10-year period and to develop a prioritized list of recommended space programs, missions, and supporting activities to address those questions. Among the seven study panels organized under that study was a water panel, whose mission would have overlapped somewhat with this study as originally envisioned.

Around the same time, the COHS was made aware that the National Science Foundation (NSF) was considering the establishment of "observatories" with field measurements and cyberinfrastructure to assemble water data into a common framework. At the time, planned observatories included the HYDRO

Program of the Consortium of Universities for the Advancement of Hydrologic Science (CUAHSI) in NSF's GEO (Geosciences) Directorate and the Collaborative Large-Scale Engineering Analysis Network for Environmental Research (CLEANER) supported by the Engineering Directorate (these two are now being combined into a program called WATer and Environmental Research Systems [WATERS] Network), along with the National Ecological Observatory Network (NEON) supported by the Biological Sciences Directorate and the Critical Zone Observatories (CZO) funded by the Directorate for Geosciences. (As of this writing, not all of these proposed activities have been funded.) And the U.S. Geological Survey (USGS), the National Oceanic and Atmospheric Administration (NOAA), the U.S. Department of Agriculture (USDA), and other agencies had been collecting various data from field measurements and experimental watersheds for many years and were exploring ways to use modern technology to increase information content and reduce costs.

Thus, while NASA, NOAA, and the USGS Earth Resources Observation Systems (EROS) Data Center were collecting satellite information, other programs were collecting field data and additional data-intensive programs were planned for the near future. But how would the collection of these data and information be optimized? How would satellite, airborne, and in-situ data be integrated? What overarching principles might be followed? This was an issue of considerable interest to NSF and, as it turned out, to other federal agencies, including NASA.

The committee met six times between February 2005 and September 2006 in open and closed sessions. Representatives from many federal, state, and nongovernmental agencies attended and participated in open sessions. This report is based on those discussions, the scientific literature, and the best professional judgment of the committee members.

The committee was deeply saddened by the September 30, 2005, passing of Leal Mertes, who was a respected and well-liked colleague and a valued committee member. We are grateful for her contributions both to this report and to our own lives.

We have many other people to thank for their help over the course of this project and in the preparation of this report. In particular, we would like to express appreciation to the following individuals for their presentations, discussions, and written submissions: Peter Arzberger, University of California, San Diego; Chaitan Baru, San Diego Supercomputer Center; Elizabeth Blood, National Science Foundation; Art Charo, NRC Space Studies Board; Tim Cohn, U.S. Geological Survey; Jared Entin, National Aeronautics and Space Administration; Frank Gehrke, California Cooperative Snow Surveys; Tom Harmon, University of California, Merced; Robert Hartman, National Weather Service; Jin Huang, National Oceanic and Atmospheric Administration; Doug James, National Science Foundation; Toshio Koike, The University of Tokoyo; Richard Lawford, GEWEX; Dennis Lettenmaier, University of Washington; Kent Lindquist, Lindquist Consulting; David Maidment, University of Texas at Austin; Robert Mason, U.S. Geological Survey;

Tom Nicholson, Nuclear Regulatory Commission; Peter van Oevelen, GEWEX; Jeff Talley, University of Notre Dame; and Stu Townsley, U.S. Army Corps of Engineers.

This report has been reviewed in draft form by individuals chosen for their diverse perspectives and technical expertise, in accordance with procedures approved by the NRC's Report Review Committee. The purpose of this independent review is to provide candid and critical comments that will assist the NRC in making its published report as sound as possible and will ensure that the report meets institutional standards for objectivity, evidence, and responsiveness to the study charge. The review comments and draft manuscript remain confidential to protect the integrity of the deliberative process.

We wish to thank the following individuals for their review of this report: Jean M. Bahr, University of Wisconsin, Madison; Matthew W. Becker, University of Buffalo; Charles T. Driscoll, Syracuse University; Dara Entekhabi, Massachusetts Institute of Technology; Tony R. Fountain, University of California, San Diego; Charles D. D. Howard, Canada; Anne W. Nolin, Steven W. Running, University of Montana; Anne W. Nolin, Oregon State University; and Claire Welty, University of Maryland, Baltimore County.

Although the reviewers listed above have provided many constructive comments and suggestions, they were not asked to endorse the conclusions or recommendations, nor did they see the final draft of the report before its release. The review of this report was overseen by Dr. Mary P. Anderson, University of Wisconsin, Madison. Appointed by the National Research Council, Dr. Anderson was responsible for making certain that an independent examination of this report was carried out in accordance with institutional procedures and that all review comments were carefully considered. Responsibility for the final content of this report rests entirely with the authoring committee and the institution.

Kenneth Potter, *Chair*
Eric Wood, *Vice Chair*

Contents

Summary

Water in the right quality and quantity, and at the right time, is essential to life—for humans and their food crops, and for ecosystems. Millions of people yearly die of water-related diseases; floods and droughts also cause illness and death in addition to economic damage throughout the world. Much of our agricultural activity would collapse in the absence of irrigation water. Natural ecosystems are adapted to stream discharge, precipitation, and evaporation patterns. Thus, future adjustments in the water cycle to climate, weather, and land-use change will undoubtedly have complex and significant impacts on humans and other species that depend on it.

THE CHALLENGE OF UNDERSTANDING WATER FLOWS AND STORES

The management of water resources to meet these challenges will require improvements in our capacity to understand and quantify the hydrologic cycle and its spatial and temporal interactions with the natural and built environment. Just as in balancing a bank account, it is important to keep track of the amount of water in storage and the rates of inflow and outflow. Natural inflows to surface-water bodies typically include precipitation, surface runoff, and groundwater inflow; outflows include evaporation, transpiration, and seepage into the ground. Natural inflow to groundwater (recharge) results from the percolation of soil and other surface waters; outflows include transpiration and discharge to surface waters. Humans also withdraw from, and discharge to, surface and groundwater.

There are two main challenges to understanding and quantifying the movement of water between and within stores and the associated changes in water constituents. First, many of the key processes, such as evaporation or movement of groundwater within an aquifer, cannot be readily observed over large areas. Second, the rates of water movement can vary greatly in both space and time.

Three strategies are used individually or in combination to get around these problems, with varying degrees of success. First, when feasible, variables such as precipitation, river discharge, and wind speed are measured at specific locations ("point" measurements). Second, remote sensing methods can provide information on the spatial distribution of key variables. These methods measure over large areas, but the measurements represent averages over some space and time "window". Remote sensing can be ground based (e.g., radar estimates of rainfall), or based on aircraft or satellites. Third, models are used to interpolate between point measurements (e.g., precipitation), estimate unmeasured quantities based on measured ones (e.g., chlorophyll concentrations from certain wavelengths of light, or evaporation from wind speed, temperature, and relative humidity) and to predict hydrological conditions under a hypothetical future combination of land use, land cover, and climate.

Even with these tools, the field suffers great limitations in many areas of measurement. For example, for most aquifers there are no accurate estimates of recharge, especially their spatial and temporal resolution. Likewise, accurate measurements of the spatial distribution of snow water storage are virtually impossible to make in many areas due to extreme topography and/or limited access.

THE PURPOSE OF THIS REPORT

Out of interest in these issues and their implications, in 2005 the National Research Council's (NRC) Water Science and Technology Board (WSTB) formed the Committee on Integrated Observations for Hydrologic and Related Sciences to examine the potential for integrating new and existing spaceborne observations with complementary airborne and ground-based observations to gain holistic understanding of hydrologic and related biogeochemical and ecological processes and to help support water and related land resource management. Funding for this effort (or for its parent standing committee, the Committee on Hydrologic Science [COHS]) was received from the National Aeronautics and Space Administration (NASA), the National Science Foundation (NSF), the U.S. Army Corps of Engineers (USACE), the National Oceanic and Atmospheric Administration (NOAA), the Nuclear Regulatory Commission, and the Environmental Protection Agency (EPA). The full Statement of Task is shown in Box S-1. This report offers a comprehensive view of the current state of integrated observing for hydrology and the related sciences, with a particular emphasis on sensing.

BOX S-1
Statement of Task

This study will examine the potential for integrating new and existing spaceborne observations with complementary airborne and ground-based observations to gain holistic understanding of hydrologic and related biogeochemical and ecological processes and to help support water and related land resource management. These systems are closely interconnected, and a great deal of common information is required in their study. The goal is to focus on information that would contribute to quantifying current and projected water availability, water quality and biogeochemical cycling, and land-surface and related ecologic conditions. The assessment would consider these goals in the light of the capabilities of sensor and other in-situ monitoring technologies and of spaceborne observation technologies. It would also look at likely advances in these technologies. The study will:

(1) Identify processes in water flow and transport, related biogeochemical cycling, and ecological impacts where better information is needed to understand important mechanisms, how systems integrate at watershed and larger scales, and where new instrumentation or strategies for instrument placement could supply the needed data;
(2) Identify contributions that observations obtained by remote sensing or other existing technology could make to understanding water flow and transport and related biogeochemical cycles as well as for addressing water management activities such as agricultural and municipal water supply, flood and drought prediction, water quality, and energy production;
(3) Evaluate the readiness of the scientific and technical communities to make effective use of more precise and reliable observations of hydrologic fluxes and states (e.g., soil moisture, snow cover, carbon and nutrient transport, water bodies and wetlands, and water-quality indicators);
(4) Suggest research opportunities in these areas; and
(5) Identify gaps in federal agency plans for integrating across sensors and products obtained from either in-situ or space-based observations.

OVERVIEW

The good news is that recent and potential future technological innovations offer unprecedented possibilities to improve our capacity to observe, understand, and manage hydrologic systems. Sensors are being developed that are smaller, less expensive, and require less power, allowing for deployment in much larger numbers. Researchers are designing sensors to provide previously unavailable information, such as real-time measurements of nutrient concentrations in surface, soil, or groundwater. Sensors are being arrayed in networks that enable the sharing of information and hence produce synergistic gains in observational capacity; these sensor networks offer the promise of filling critical gaps between traditional point and remotely sensed measurements. New sensors are being deployed on aircraft and satellites, and new ways are being thought of to use existing remote sensors. Computer models are being used to assimilate data from multiple sources to predict system behavior. And cyberinfrastructure initiatives are providing efficient and effective ways to share data with scientists, managers, and other potential users.

But there are gaps between the vision of what researchers and managers want to achieve and their ability to realize that vision. These gaps are real, but in many cases extremely narrow. Technical challenges include those associated with the development of robust, accurate, and affordable water-quality sensors. All of the required technologies are expensive to develop, and most will at least initially require public funding until their commercial viability is established. Design and implementation of integrated hydrologic measurement systems requires the cooperation of diverse sets of researchers, technologists, and decisionmakers. How can this cooperation be facilitated? Water management in the United States is typically local and rarely integrated. How can integrated hydrologic measurements provide greater benefit to local or regional decisionmaking?

A series of case studies was developed for the report, for regions as diverse as the Arctic, sub-Saharan Africa, and the Everglades, as well as the Great Plains, Mountain West, and coastal North Carolina of the United States.

- "Monitoring the Hydrology of the Everglades in South Florida" provides an example of a large, complex integrated observatory designed to address pressing water management needs in an ecologically sensitive area.
- "Impacts of Agriculture on Water Resources: Tradeoffs between Water Quantity and Quality in the Southern High Plains" focuses on semiarid regions where water availability is a critical issue and where cycling of salts has large-scale impacts on water quality.
- "Hydrological Observation Networks for Multidisciplinary Analysis: Water and Malaria in Sub-Saharan Africa" illustrates how observations can contribute to understanding and ameliorating major water-related public health problems.
- "Achieving Predictive Capabilities in Arctic Land-Surface Hydrology" explores a strategy for robust remote sensing hydrology in the pan-Arctic, to identify capabilities needed to link in-situ observations to satellite sensor-scale observations.
- "Integrating Hydroclimate Variability and Water Quality in the Neuse River (North Carolina, USA) Basin and Estuary" focuses on the impact of human activity and hydroclimate variability on watershed nitrogen sources, cycling and export, and consequently on fresh water and estuarine ecosystem health.

- Finally, "Mountain Hydrology in the Western United States" discusses the need for high spatio-temporal resolution information on snow conditions due to sharp wet-dry seasonal transitions, complex topographic and landscape patterns, steep gradients in temperature and precipitation with elevation, and high interannual variability.

These case studies, taken in the context of discussions of sensors, networks, communications, data assimilation, and modeling, illustrate a number of important challenges regarding current and potential sensors and sensor networks, merging the resulting information with models, and providing useful products to managers and policymakers through traditional and emerging dissemination media. These challenges are presented below.

CHALLENGES

Development and Field Deployment of Land-Based Chemical and Biological Sensors

Physical sensors, such as those that measure air and water temperature and pressure, radiation, and wind speed and direction, are now mass produced and routinely packaged together in small instruments along with power and communication devices. However, sensor development for many important chemical and biological measurements is relatively immature. Development of a wide range of field-robust chemical and biological sensors is one of the greatest challenges facing widespread deployment of sensor networks in the hydrologic sciences.

Airborne Sensors

Airborne measurements operate at a spatial scale that fills the gap between the in-situ plot-scale observations and the larger satellite-scale observations. Airborne remote sensing at NASA historically was viewed as an intermediate step between initial sensor development and space deployment to help develop retrieval algorithms to validate new satellite sensors. It has not been viewed as a sensor program in its own right. This has impeded the development of operational airborne observing platforms that could play a very important role in hydrologic observations.

Spaceborne Sensors

In satellite-based remote sensing, NASA has made good progress in developing and deploying sensors used primarily for research. Nonetheless, two challenges are relevant to this report: (1) a resolution of the "research-to-operations" transition from NASA-developed "experimental" satellite observations to the broad variety of operational agencies and users that need routine (i.e., operational) observations, and (2) the lack of a corresponding monitoring strategy by entities such as EPA, USDA, NOAA, and state water and natural resources agencies that would incorporate airborne and/or satellite remote sensing measurements, where appropriate.

Bridging the Gap between Sensor Demonstration and Integrated Field Demonstration

There are significant interagency gaps between the steps of sensor development, sensor demonstration, integrated field demonstrations, and operational deployment of sensors. The greatest gap is between sensor demonstration and integrated field demonstration. Closing this gap would involve integrating the sensor networks and webs within hydrologic observatories and experimental demonstration sites, and interfacing the sensor networks with the broader development of cyberinfrastructure.

Integrating Data and Models for Operational Use

The importance of data-model integration is apparent in a number of the case studies. For the Mountain Hydrology study, predictions of water availability are made from point measurements and model forecasts. In the Neuse River Basin study, management decisions are based on sparse water-quality measurements. In each case, models and observations are used to guide management decisions, and in each case a data assimilation system that merges models and observations would offer improved predictions. The challenge is to develop methods that will be useful for broad families of applications, rather than just a few of the many possible applications.

The Next Step: Water Resources Applications

In the United States, large water resources problems involve multiple stakeholders, including government agencies, business interests, and the public. Management is typically diffuse, and standard measurement and modeling techniques and rules for water management are entrenched and often legally mandated. This produces a consistent data set to show trends over time, and simplifies training and daily tasks of staff. However, it also leads to missed opportunities to improve the accuracy and precision of the data and resulting model predictions.

Funding Highly Interdisciplinary Science

Interdisciplinary science is increasingly common, but the design and use of integrated hydrologic measurement systems in specific research applications adds complexity to the challenge. These new kinds of projects will require unprecedented interdisciplinary cooperation among electrical engineers, computer scientists, modelers, and the physical, chemical, and biological scientists who apply technology to hydrologic research. While many universities and research laboratories have the required expertise, marshalling this expertise on specific projects will likely require new programs or sources of funding.

Addressing the Fractured Federal Responsibility for Hydrologic Measurement, Monitoring, and Modeling

The overarching barrier to the development and implementation of integrated hydrologic measurement systems is the lack of a single federal agency with primary responsibility for measuring, monitoring, and modeling the environmental factors and processes that control the hydrologic cycle. It is easy to understand why the responsibility for measuring and monitoring the environmental factors and processes that control the hydrologic cycle might have evolved as it has. But the dual threats of global climate change and population growth demand a focused strategy for providing information on the nation's water resources and the environment.

The above challenges, along with the body of the report, lead to the recommendations of the study, which follow below.

RECOMMENDATIONS

Recommendation 1: NSF, in partnership with NASA, NOAA, EPA, U.S. Geological Survey, and possibly national health and security agencies, and with collaboration from the private sector, should develop one or more programs that address the need for multidisciplinary sensor development. An interagency sensor laboratory should be considered.

Recommendation 2: Serious consideration should be given to empowering an existing federal agency with the responsibility for integrated measurement, monitoring, and modeling of the hydrological, biogeochemical, and other ecosystem-related conditions and processes affecting our Nation's water resources.

Recommendation 3: Coordinated and jointly funded opportunities for observatories, demonstration projects, test beds, and field campaigns should be significantly increased.

Recommendation 4: Agencies should consider offering new funding streams for projects at the scale of several million dollars per year for approximately 5-10 years to help close the gap between sensor demonstration and integrated field demonstration.

Recommendation 5: NASA should strengthen its program in sensor technology research and development, including piloted and unpiloted airborne sensor deployment for testing new sensors and as a platform for collecting and transmitting data useful for applications.

Recommendation 6: In addition to partnerships with other federal agencies for the development and testing of experimental sensors that are of a particular interest to agencies, the Nation, and especially NASA, should explore additional strategic partnerships with space agencies in other countries and regions, such as the European Space Agency (ESA), the Japanese Aerospace Exploration Agency (JAXA), the Centre National d'Études Spatiales (France), and the Canadian Space Agency (CSA).

Recommendation 7: NASA and NOAA should work with NSF and other agencies to assure that plans for incorporation of space-based and airborne observations (from both existing and, preferably, planned or proposed missions) are part and parcel of the experimental design of these proposed observatories.

Recommendation 8: Advanced cyberinfrastructure should not only be incorporated as part of planned observatories and related initiatives to help manage, understand, and use diverse data sets, but should be a central component in their planning and design.

Recommendation 9: Utilization of web-based services, such as collaboratories (i.e., web-based systems where researchers and users come together to build a system of data, predictive models, and management projects), for the distribution of observations, model predictions, and related products to potential users, should be encouraged.

Recommendation 10: NASA and NSF should develop and strengthen program elements focused on demonstration projects and application of data assimilation in operational settings where researchers work collaboratively with operational agencies.

Recommendation 11: NASA should take the lead by expanding support for the application of integrated satellite remote sensing data products. NSF, NOAA, and other federal and state agencies engaged in environmental sensing should likewise expand support for the creation of the integrated digital products that meet educational, modeling, and decision-support needs.

Recommendation 12: Congress, through the budgetary process, should develop a strategy for transitioning NASA experimental satellite sensors to operational systems with assured data continuity so that the Nation's investment in remote sensing can be utilized over the long term by other federal agencies and users.

Recommendation 13: Water agencies should be alert for opportunities to incorporate new sensor and modeling technologies that will allow them to better deliver their mission and be more productive.

1

Introduction

THE CHALLENGE

Water is essential to life; human well-being requires an adequate and continuous supply of clean water. As of 2004, about one-sixth of the world's population (over 1 billion people) remained without access to improved drinking water, and over 40 percent (about 2.6 billion) had no access to improved sanitation (World Health Organization and UNICEF, 2006). Although the United States is blessed with abundant water, there are many regions with intermittent and even chronic water shortages. As populations increase, water problems will likely worsen. Land-use changes associated with population growth commonly increase flood peaks, lower base flows, and impair water quality. As worldwide stores of oil and gas become depleted, increased amounts of water may be required for biofuels production, coal gasification, and processing of oil shale and other alternative fuels. Anthropogenic and natural climate change will compound the problem by altering historical climate and weather patterns in unpredictable ways. And human demands for water are increasingly coming into conflict with environmental water use. There is little doubt that the future offers many challenges to the maintenance of adequate supplies of clean water.

The science questions associated with these challenges have been summarized in a series of reports by the National Research Council from the early 1990s to 2007; these are collated in Appendix A. These themes are mostly echoed in the science plans of the National Science Foundation's proposed ecological, hydrologic, and water-quality observatories, reflecting considerable consensus in the scientific community. Progress in answering these questions has been disappointingly modest, due in part to the limited availability and/or use of the technologies and strategies described in this report. Because the societal issues and associated science questions are evolving only gradually, this report does not attempt a detailed recapitulation of these topics. However, they appear throughout this chapter and, indeed, throughout the report.

What is clear is that to meet these challenges the management of water resources will need to become more effective and systematic (see, for example, an international example in Box 1-1). This will require improvements in our capacity to understand and quantify the hydrologic cycle and its spatial and temporal interactions with the natural and built environment.

USES OF WATER AND THE HYDROLOGIC CYCLE

Water on the Earth continuously moves between the atmosphere, oceans, and various water stores on and below the land surface (Figure 1-1). The primary land-based water stores, in order of capacity, are icecaps and glaciers, groundwater, lakes, terrestrial soils, wetlands, and rivers. Water use and management generally focus on groundwater, lakes, rivers, and soils. These stores provide water for domestic, commercial, industrial, and agricultural uses. They have also been used to dispose various kinds of wastewater, although such uses have become increasingly constrained.

Ecosystems rely on and participate in the cycling of water. Water supports most forms of life and water stores are often associated with characteristic ecosystems. Late in the 20th century, societies recognized the utility of protecting and restoring ecosystems to maintain a "healthy" hydrologic cycle. For example, wet-

BOX 1-1
An International Initiative for Integrated Observations

The international community is recognizing the importance of integrated, comprehensive Earth monitoring systems. Perhaps most notably, on February 16, 2005, at the Third Earth Observation Summit held in Brussels, Belgium, the Global Earth Observation System of Systems (GEOSS) 10-Year Implementation Plan was endorsed by nearly 60 governments and the European Commission. The vision for GEOSS is "to realize a future wherein decisions and actions for the benefit of humankind are informed via coordinated, comprehensive and sustained Earth observations and information." The GEOSS implementation plan calls for developing these comprehensive Earth observations in order to improve monitoring the state of the Earth, to increase understanding of Earth processes, and to enhance prediction of the behavior of the Earth system. The ultimate goal is to deliver to Earth's societies enhanced benefits from improved forecasting and management in areas such as weather and seasonal climate, water resources and ecosystem management, environmental factors affecting human health and understanding, climate variability and change. The intergovernmental Group on Earth Observations (GEO), which is leading the effort, is co-chaired by the United States, South Africa, China, and the European Commission.

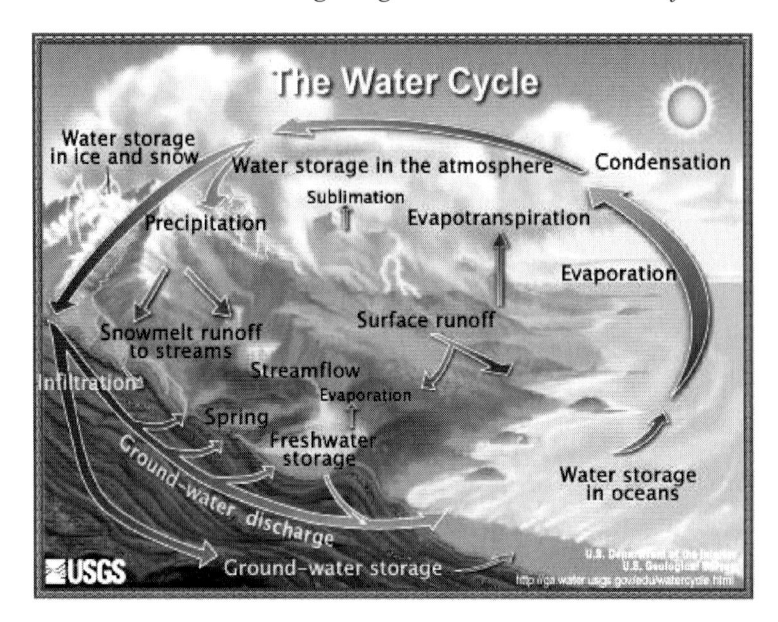

FIGURE 1-1 The water cycle. SOURCE: U.S. Geological Survey. Available on-line: http://ga.water.usgs.gov/edu/watercycle.html.

wetlands, many of which were drained for agricultural and other land uses, are now protected and restored in recognition of their role in moderating floods, maintaining dry-weather flows, and removing sediment and overabundant nutrients (NRC, 1995).

Water continuously moves between and within water stores. The rates of movement between stores, and consequently the amount of water stored, can vary significantly, posing challenges to water management. Droughts compromise water use and floods jeopardize activities sited near water bodies. The "quality" of water depends on its constituents, which include heat, sediment, nutrients, carbon, and harmful anthropogenic materials (contaminants). As water moves between and within stores, its constituents change. Some constituents, such as quartz sand, change only slowly during transport (Selley, 2000), although transport may be interrupted along the way. Other constituents undergo major transformations. For example, under certain conditions nitrate is converted to nitrogen gas as it moves from groundwater to surface water. Transformations typically involve a combination of physical, chemical, and biological processes. Effective water management requires that we understand and quantify the movement of water between and within stores, and the movement and transformation of important water constituents.

Of primary importance are the water budgets of important water stores and their variation over time. Just as in balancing monetary accounts, it is important to keep track of the amount of water in storage as well as the rates of inflow and outflow. Natural inflows to surface-water bodies typically include precipitation, surface runoff, and groundwater inflow; outflows include evaporation, transpiration, and seepage into the ground. Natural inflow to groundwater (recharge) results from the percolation of soil and other surface waters; outflows include transpiration and discharge to surface waters. Humans also withdraw from, and discharge to, surface and groundwater; the sum of the natural and human infrastructure systems can lead to highly complicated water flow patterns and budgets (e.g., Figure 1-2).

Some aspects of global, regional, and local water budgets are better understood than others (Trenberth et al., 2007). For example, water accounting for lakes is relatively easy, although it can be difficult to obtain accurate estimates of evaporation and exchanges with groundwater. In contrast, it is often very difficult to estimate groundwater stores. Groundwater levels vary in space, and there are generally relatively few wells from which to make observations. Groundwater recharge depends on precipitation and evaporation, both of which are difficult to measure over large areas. For many important ground water stores (aquifers), pumping rates are not well known.

As previously mentioned, the quality of water depends on its constituents, which in turn are affected by the pathways and rates of water movement within water stores. Of particular importance is the time that individual constituent units (such as particles of sediment) spend in a store (i.e., the residence time). For example, the percentage of suspended materials carried into a quiet lake by streamflow that will settle out depends in part on the residence time, in accordance with Stokes' Law, and its variants (Wu and Wang, 2006). In contrast, a long residence time in a lake may concentrate the chemical constituents by evaporation (Warren, 2006). Bacteria accumulate at different rates under certain conditions than under others. Hence to accurately predict the quality of water leaving a store it is necessary to understand and quantify the movement of water within the store. This can be much more difficult than estimating water budgets.

QUANTIFYING THE HYDROLOGIC CYCLE

There are two main challenges to understanding and quantifying the movement of water between and within stores and the associated changes in water constituents. First, many of the key processes cannot be readily observed. Evaporation is a good example. Movement of groundwater within an aquifer is another. Second, the rates of water movement can vary greatly in both space and time. For example, groundwater recharge depends on precipitation, soil properties, topography, and vegetation, all of which can vary widely in space and to some extent with time.

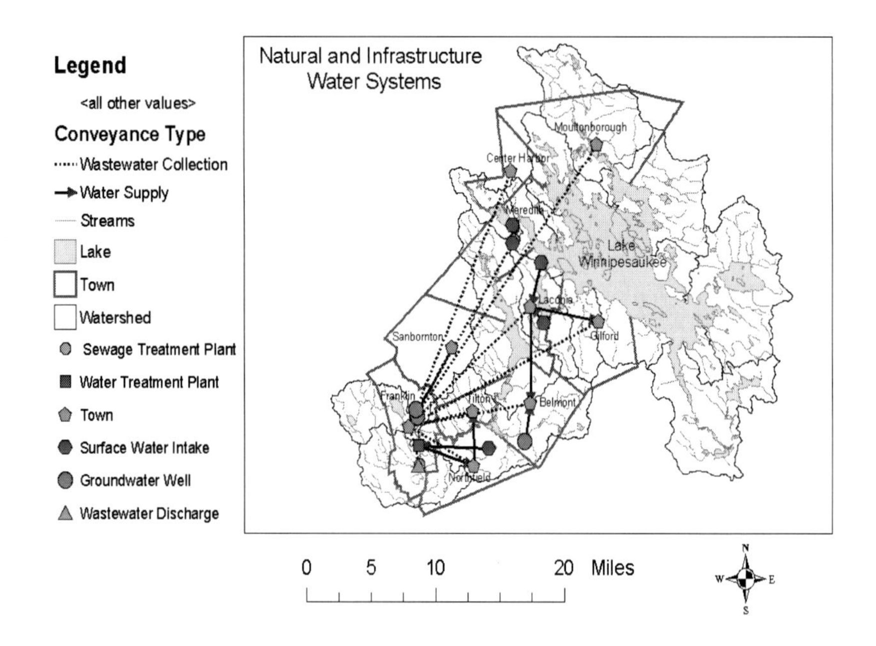

FIGURE 1-2 Combination of the natural and infrastructure water systems, Lake Winnipesaukee, central New Hampshire. The figure demonstrates the complexity of the modern water cycle when water use is superimposed on the natural system. SOURCE: NRC (2002).

Three strategies are used individually or in combination to get around these problems, with varying degrees of success. First, when feasible, variables are measured at specific locations ("point" measurements). These include key terms in the water budget, such as precipitation, as well as other variables that correlate with water budget terms. Examples of the later include river stage, which is used to estimate river discharge; and air and water temperature, wind speed, and relative humidity, which are sometimes used to estimate lake evaporation.
Second, remote sensing methods are used to provide information on the spatial distribution of key variables. These methods measure over large areas, but the measurements represent averages over some space and time "window". For example, radar estimates of rainfall usually correspond to an area of 1-4 km^2 and a time interval of 5-10 minutes. Remote sensing can be ground based, as in the case of radar measurements of rainfall, or based on aircraft or satellites. In general, the size of the space and time window increases with the elevation of the sensor.

Third, models are used to extend the utility of measurements. Models can be used to interpolate between point measurements. For example, models are used to estimate spatially distributed precipitation in mountainous watersheds based on point measurements and topography. Models can be used to estimate unmeasured quantities based on measured ones. Some such models are based on statistical relationships. For example, statistical models are used to estimate chlorophyll concentrations based on satellite measurements of certain wavelengths of light (Bailey and Werdell, 2006). Other models are based on physical principles. The estimation of lake evaporation from wind speed, temperature, and relative humidity is an example, although in practice it is usually necessary to calibrate the model using independent estimates of evaporation. Finally, models can be used to predict hydrological conditions under hypothetical conditions, such as future land use, land cover, and climate.

SHORTCOMINGS OF TRADITIONAL MEASUREMENT TECHNIQUES

Point measurements, remotely sensed measurements, and models have been used effectively to manage water resources and conduct research on hydrologic processes. However, traditional approaches have serious shortcomings. Even using models to combine point and remotely sensed measurements, it is often not possible to accurately characterize the spatial and temporal variation of critical water budget components, such as precipitation, evaporation, and ground water recharge. In many applications it is not even possible to obtain accurate estimates of the temporally and spatially averaged components, and hence "close the budget". For example, for most aquifers there are no accurate estimates of recharge. Finally, currently used methods rarely provide sufficient information on the movement of water within important water stores, such as groundwater and lakes.

These failings are particularly problematic given the challenges of population growth, land-use change, climate change, and the growing importance of water and environmental quality. This is illustrated below in the context of surface-water quality and groundwater recharge.

The Federal Water Pollution Control Act (33 U.S.C. 1251 et seq.), commonly known as the Clean Water Act, requires that states identify sources of pollution in waters that fail to meet state water-quality standards and develop strategies to meet the standards. This process is often based on the use of water-quality models that are calibrated using historical water-quality data (e.g., Soil and Water Assessment Tool [SWAT], http://www. brc.tamus.edu/swat/). For the most part these models grossly simplify the processes that control the quality of natural waters. Water quality may vary greatly in space and time based on a large number of variables. A short list of these variables would include natural factors such as geology, topography, meteorology, hydrology, and biology; and human activities such as urbanization, agriculture, and industrial activity. Seasonal variations are common, and even within a single storm event the "first flush" may contain a cocktail of hydrocarbons, heavy metals, and nutrients, whereas the later flow may be comparably clean (NRC, 1994). Given the expense of traditional water-quality data collection, it should be no surprise that few projects are able to resolve these variations.

Consider, for example, the disposition of sediment, phosphorus, nitrogen, and carbon as these constituents move through a watershed. Although there is a general understanding of the various processes involved, there is little quantitative information for specific watersheds. What is the residence time of sediment particles and associated phosphorus at key storage sites, such as colluvial deposits, floodplains, and wetlands? What is the bioavailability of that phosphorus. and how does it change over time? Where does denitrification occur and at what rate? How would changes in land use, land cover, and climate affect the transport of sediment, nutrients, and carbon? What is the most cost-effective configuration of management practices? The goals of the Clean Water Act may be difficult to meet until these questions are answered at the required level of accuracy for specific watersheds, or at least for appropriately defined classes of watershed.

Similar issues exist with respect to groundwater. Basin-wide groundwater recharge for watersheds with stream gages is often estimated from the streamflow records by separating out the baseflow component from the stormflow component using a variety of methods, or by evaluating displacement in the recession curve after a storm. But these approaches generally do not provide information on the horizontal distribution of recharge or on vertical partitioning in the case of multiple aquifers. Point estimates of recharge can be made by a variety of methods based on temperature, solutes, isotopes, and moisture contents, and water levels, but their utility can be limited due to high spatial variability. Rainfall-runoff models provide some information on the spatial distribution of recharge, but are limited by uncertain information on the spatial distribution of precipitation, evapotranspiration, and soil properties, as well as by limited understanding of the recharge process.

One critical knowledge gap is the role of focused recharge vs. diffuse recharge. It is well known (Logan and Rudolph, 1997; Hayashi et al., 2003) that focused recharge via surface-water bodies dominates in some landscapes; but there is also evidence that focused recharge through small depressions is important in other landscapes (Delin et al., 2000). Modeling focused recharge requires accurate, high-resolution topographic information, which is just becoming available through technologies such as LiDAR. Groundwater models are also useful for estimating the spatial distribution of recharge, although such estimation requires many observation wells and the results are generally confounded by spatial variation in aquifer properties.

A VISION FOR THE FUTURE

The looming challenges of population growth and global climate change demand that we improve our capacity to observe, understand, and manage hydrologic systems. Fortunately, recent and potential future technological innovations offer unprecedented possibilities to do so. Sensors are being developed that are smaller, less expensive, and require less power, allowing for deployment in much larger numbers. Researchers are designing sensors to provide previously unavailable information, such as real-time measurements of nutrient concentrations in surface, soil, or groundwater. Sensors are being arrayed in networks that enable the sharing of information and hence produce synergistic gains in observational capacity; these sensor networks offer the promise of filling critical gaps between traditional point and remotely sensed measurements. New sensors are being deployed on aircraft and satellites, and new ways are being thought of to use existing remote sensors. Computer models are being used to assimilate data from multiple sources to predict system behavior. And cyberinfrastructure initiatives are providing efficient and effective ways to share data with scientists, managers, and other potential users. An integrated hydrologic measurement system that exploits these important innovations would significantly increase our capacity to understand and manage our water resources.

Imagine, for example, the extent to which environmental forecasting and management would have been improved during the fall of 1999 when Hurricanes Dennis, Floyd, and Irene impacted North Carolina if there had existed an integrated hydrologic meas urement system. The combined use of remote sensing, soil moisture sensor networks, and spatially distributed hydrologic modeling would have provided better estimates of runoff from spatially and temporally variable source areas. The use of recently obtained statewide LiDAR elevation data would have enabled more accurate forecasting of floodwater levels in streams, rivers, and coastal areas. The coupled use of emerging biochemical nutrient sensors, soil moisture measurement networks, and an expanded, real-time, stream water quality sensor network would have provided estimates of the changes in nutrient stores and fluxes associated with storm and inter-storm periods. Use of all of the above-mentioned data in advanced atmospheric, hydrodynamic, and ecosystem models would have enabled the forecasting of nutrient flushing, water quality, and long-term impacts on the riverine and estuarine ecosystems, and would have likely provided critical insights into coupled system behavior. Finally, an advanced environmental cyberinfrastructure representation of the watershed based on the observational and modeling data would have provided the information required to assess vulnerability and risk to populations in different parts of the basin and coordinate emergency response. As another example, consider drought. In the United States, the persistence of low precipitation in the West led to a series of Western Governors' Association meetings calling for a more effective drought early warning system (http://www.westgov.

org/wga/publicat/nidis.pdf). This resulted in the creation of the National Integrated Drought Information System (NIDIS), with the National Oceanic and Atmospheric Administration (NOAA) being the lead agency. Central to agency plans for drought information is an attempt to strengthen observing systems, integrate operational systems that exist at federal, state, and local levels, combine this information with satellite observations and forecast weather models operated by NOAA, and to deliver the information and early warning through a "web portal". An effective NIDIS information system will require smart sensors to measure snow pack and melt—critical springtime data for the West—and to measure summertime soil moisture and near-surface meteorology (e.g., humidity), as well as satellite information that can provide a spatial perspective to the local in-situ measurements.

However, before these societal benefits can be reaped, there are significant challenges to overcome. Major technical challenges include those associated with the development of robust, accurate, and affordable water-quality sensors. All of the required technologies are expensive to develop, and many if not most will at least initially require public funding. Design and implementation of integrated hydrologic measurement systems require the cooperation of diverse sets of researchers, technologists, and decisionmakers. How can this cooperation be facilitated? Water management in the United States is typically local and rarely integrated. How can integrated hydrologic measurement benefit local decisionmaking? Or will the problems of population growth and climate argue for integrated water-resource decisionmaking (Potter, 2006)? This report explores these and other issues.

SCOPE AND ORGANIZATION OF THIS REPORT

The NRC's Committee on Integrated Observations for Hydrologic and Related Sci-ences was asked to provide guidance and advice to its sponsors in accordance with its statement of task (Box 1-3). This report contains the results of that study.

The global water cycle and the corresponding international water programs serve as the backdrop for this report, and international themes are echoed throughout it. These are illustrated in Boxes 1-1 and 1-2 of this chapter, in the Central America Flash Flood Guidance System example of Chapter 3, in the case studies on "Water and Malaria in sub-Saharan Africa" and "Achieving Predictive Capabilities in Arctic Land-Surface Hydrology" of Chapter 4, and even in the recommendation for international collaboration in Chapter 5. However, consistent with the statement of task and the report title, the study emphasizes the U.S. experience. The study was requested, and funded, by U.S. agencies. Further, understanding and integrating the roles of the major international monitoring and observing systems would have added a level of complexity to the study

BOX 1-2
Special Challenges in Developing Countries: An East African Example

Drought in less-developed regions poses additional observational problems to overcome if the promise of integrated observation systems is to be realized. In East Africa, drought has become severe after below-average rainfall in the first six years of the 21st century. In this region, there typically is a long, rainy season from February to June and a short, rainy season October to December. In 2005, initially good rains in January were not sustained; this resulted in drought that became severe when the fall, rainy season failed to materialize, with annual rainfall totals of 20 to 60 percent of normal (NOAA Climate Prediction Center). Interspersed with this lack of rainfall were devastating floods in the April to June year period in Ethiopia, Somalia, Kenya, and Uganda that left almost 200 dead and 300,000 displaced (Dartmouth Flood Observatory, http://www.dart-mouth.edu/~floods).

The sparseness of in-situ meteorological hydrological networks has hampered the monitoring of flood and drought, especially the severity and extent, at sufficiently high resolution to be useful for disaster relief. Additionally, the development of more-skillful forecast models, ranging from short-term weather models to seasonal climatic and hydrological forecasting requires denser in-situ observations and/or the merging of in-situ and satellite measurements of rainfall, wind, and atmospheric profiles of temperature and humidity. Quantitative satellite observations, when merged with sparse in-situ measurements, can provide areal coverage of terrestrial water and environmental variables useful for forecasting and decisionmaking.

Monitoring of weather and climate over Africa, and the assessment of drought and flood conditions is being done through the Famine Early Warning System Network funded by the U.S. Agency for International Development and operationally implemented by National Weather Service's Climate Prediction Center. The National Aeronautics and Space Administration's Earth Observatory and Natural Hazards (http://earth observatory.nasa.gov/NaturalHazards/) offers imagery related to floods, droughts, and other hazards globally. But, the above information tends to be qualitative (satellite imagery), poorly verified (e.g., the skill in seasonal precipitation and temperature forecasts), and dispersed among agencies and data centers, all making its usefulness for decisionmaking problematic. There is an information roadblock preventing the integration of these data with in-situ measurements—some technological and some institutional. An integrated Earth observation system useful for informed decisionmaking in Africa, or indeed anywhere, needs to combine new sensor technology for in-situ measurements, comprehensive satellite observations of the terrestrial water cycle, communications and data networking for the sensor networks and satellites systems, and computational platforms for merging the diverse data within models. It is also important to involve local communities and governments, taking into consideration their unique needs and customs, to ensure that equipment is maintained and human observations are integrated with sensor data.

BOX 1-3
Statement of Task

This study will examine the potential for integrating new and existing spaceborne observations with complementary airborne and ground-based observations to gain holistic understanding of hydrologic and related biogeochemical and ecological processes and to help support water and related land resource management. These systems are closely interconnected, and a great deal of common information is required in their study. The goal is to focus on information that would contribute to quantifying current and projected water availability, water quality and biogeochemical cycling, and land-surface and related ecologic conditions. The assessment would consider these goals in the light of the capabilities of sensor and other in-situ monitoring technologies and of spaceborne observation technologies. It would also look at likely advances in these technologies. The study will

(1) Identify processes in water flow and transport, related biogeochemical cycling, and ecological impacts where better information is needed to understand important mechanisms, how systems integrate at watershed and larger scales, and where new instrumentation or strategies for instrument placement could supply the needed data;
(2) Identify contributions that observations obtained by remote sensing or other existing technology could make to understanding water flow and transport and related biogeochemical cycles as well as for addressing water management activities such as agricultural and municipal water supply, flood and drought prediction, water quality, and energy production;
(3) Evaluate the readiness of the scientific and technical communities to make effective use of more precise and reliable observations of hydrologic fluxes and states (e.g., soil moisture, snow cover, carbon and nutrient transport, water bodies and wetlands, and water-quality indicators);
(4) suggest research opportunities in these areas; and
(5) identify gaps in federal agency plans for integrating across sensors and products obtained from either in-situ or space-based observations.

that would have detracted from its core messages, which are directed toward the U.S. government.

Not all technologies and topics related to water are covered in the report. For example, despite their fundamental importance to the conduct of water research at any but the smallest scale, there is only cursory mention of stream gages. This is due primarily to the fact that such gages are based on technology that was developed in the late 19th and early 20th centuries. As another example, glaciers are covered only marginally in the report. This is for two reasons. The first is that relative to related phenomena such as seasonal snowmelt, glaciers play only a small role in the water management picture in the United States. A second is that glaciers are traditionally considered part of the cryosphere—along with ice sheets, sea ice, and permafrost—and are thoroughly covered in studies involving global climate change and cold region processes.

The planning, design, operation, and utilization of an integrated observational-modeling system involves many elements, or stages. These include (a) defining goals, which may include specific "deliverables" for a narrowly defined research project or flexible targets when the project is established for broader and potentially changing uses; (b) building a team with appropriate expertise to define and oversee accomplishment of the goals; (c) designing the project to achieve the goals, either specifically or with flexibility to allow for multiple-use data; (d) collecting and validating the data, integrating and validating new data collection methods as appropriate over time; (e) organizing the large data sets for a variety of different uses; (f) integrating observations across sensors and networks; (g) merging the integrated observations with models and model validation; and (h) delivering the information products to those applying them to flood and drought forecasting, water management planning, disaster response, source water protection, and other areas. These steps are described in more detail in Appendix C.

The first three steps are explicitly or implicitly part of the case studies summarized in Chapter 4, but since they are common to all interdisciplinary projects are not discussed explicitly in the report. The last five steps are the subject of Chapters 2 and 3. Thus, Chapter 2 discusses innovations in sensor technologies. Chapter 3 outlines how data collected using existing and emerging technologies can be integrated and assimilated into models and communicated to the user. Chapter 4 uses case studies to illustrate how these innovations are, and could be, applied in specific settings. Finally, Chapter 5 synthesizes the lessons learned from the case studies and from other ongoing activities, and summarizes the committee's findings and recommendations.

2

Sensing from the Molecular to the Global Scale: New Opportunities and Challenges

The hydrologic sciences are based on data, and much of that data has to be collected using measurements of the real world, from remote locations in Africa or the Arctic to sewage systems to national forests during hunting season. Each data point results from some form of environmental sensing. Thus, the hydrological community has focused tremendous effort, in collaboration with scientists and engineers from a host of disciplines, on developing cost-effective, high-quality, reliable sensing devices from simple staff gages to complex satellite technologies coupled with geographic information systems (GIS). In the case studies provided in Chapter 4, examples are given of how a wide variety of sensing technologies and approaches can be used not only for research but also for applications such as water management and human health, on which societies depend. Rather than provide a lengthy review of the development and deployment of sensor technologies established in the hydrological sciences, this chapter focuses primarily on new opportunities and their attendant challenges for sensing of hydrologic and related parameters. The intent is to review current and anticipated capabilities for sensing from the molecular to the regional and global scales, in part to allow the reader to better understand the state-of-the-art and in part to indicate where integrative development needs to be fostered so that the potential of new sensing and information technologies can be realized.

Taken as a whole, this chapter should provide a convincing argument that the hydrological sciences are poised for a major advance brought about by the convergence of new sensors and sensing approaches from the molecular/nano- to the global scale. Because all of these new or potential technologies require interdisciplinary cooperation to achieve their full potential, it is crucial that researchers in the hydrological community communicate to those in other disciplines (e.g., electrical engineering, chemical engineering, computer science, nanotechnology) the tremendous needs and challenges in water, and their important role in addressing them.

This chapter is divided into four main sections. First, current and emerging sensor networking technologies are described in detail, with a focus on embedded sensor networks, which will provide a platform or ground-truthing for many of the other methods. Second, recent and emerging biogeochemical sensor technologies, many of which can be integrated into emerging in-situ embedded sensor networks to broaden the parameters measured at low cost, are described. Third, current capabilities and potential for Earth observations from airborne platforms are discussed. Finally, current capabilities and potential for Earth observations from spaceborne platforms are reviewed.

Although the chapter is organized in this manner, the different technologies tend to work best when they are well integrated across scales, using information from one scale to help refine strategies at another scale and purposely focusing efforts for data collection at different scales on specific locations of interest, with appropriate time and density of sampling. Temperature is a key example. It has numerous applications, such as industrial water management, aquatic habitat mapping, and crop evapotranspiration estimates. It is also a quantity that can be measured using in-situ, airborne, and spaceborne platforms.

IN-SITU SENSOR AND SENSOR NETWORKING TECHNOLOGIES

Hydrologic science has used networks of physical and, in many cases, chemical sensors for decades. Examples include the U.S. Geological Survey (USGS) stream gaging network, the Natural Resources Conservation Service (NRCS) snow telemetry network and more recently their soil moisture Soil Climate Analysis Network (SCAN), and the National Oceanic and Atmospheric Administration (NOAA) cooperative weather station network. Besides these operational networks, there is a wide range of state, local, and research networks that include state departments of transportation that monitor weather impacts on highways, state environmental and agricultural services, and research networks such as the Ameriflux network for monitoring carbon and water fluxes. The hydrologic research community has just begun to take advantage of recent developments in sensor technologies, wireless communications, and cyberinfrastructure to develop increasingly sophisticated sensor networks allowing for sampling at greater spatiotemporal resolution and for more comprehensive 'sensor-to-scientist' operation (e.g., Barrenetxea, 2006; Cayan et al., 2003; Hanson et al., 2003; Hamilton et al., 2007; Harmon et al., 2007; Seders et al., 2007).

State-of-the-art sensing capabilities for environmental observatories reflect the co-evolution of sensors (NSF, 2005), communication technologies (Porter et al., 2005), and cyberinfrastructure (Estrin et al., 2003). There are both promising opportunities and major challenges associated with effectively linking space-based and ground-based environmental observations. This section contains (1) a description of a variety of sensor network configurations that are emerging, (2) a description of the emerging sensor and communication technologies that lie at the heart of these observational systems, and (3) identification of some of the challenges to implementing these new networking approaches and technologies in actual integrated field observatories.

Sensor Network Configurations

Beyond the traditional sensor network, an emerging technology that is becoming increasingly important is the "*embedded sensor network*" (ESN; e.g., Seders et al., 2007), also referred to with a slightly different acronym as "embedded network sensing" (ENS; e.g., Harmon et al., 2007). As used in this report, an ESN is made up of spatially distributed sensor-containing platforms or *pods*, connected to and often controlled by computers, used to measure conditions at different locations, such as air temperature, solar radiation, relative humidity, or water properties. Ideally, these devices are small and inexpensive, so that they can be produced and deployed at greater density than allowed by more traditional and expensive devices, potentially in large numbers for some applications. In order to be field deployable at low cost, their energy requirements, memory, computational speed, and bandwidth (due at least in part to Federal Communications Commission (FCC) regulations) are constrained. The additional term "*sensor web*" (Delin et al., 2005) is used specifically to imply a very dense network of numerous sensors forming a high-resolution mesh; here the single term ESN is used for ease of discussion.

In an ESN (Figure 2-1 and Table 2-1), there is computer-based dialog amongst the various sensor pods and with a gateway computer. Hence, the network *embeds* a computational intelligence in the environment that allows adaptive monitoring as well as potential control of a local environment. One of the primary advantages of an ESN over a group of individual, unconnected sensors is the ability for the network itself to make decisions based on environmental indicators, such as to change sampling strategy to minimize energy requirements and extend sensor lifetime (e.g., Seders et al., 2007; Ruggaber et al., 2007). Such a network also permits real-time sensor-to-scientist operation, such as sensor calibration, tools for data mining, modeling capabilities, and data visualization (Hamilton et al., 2007). For example, sensor networks may be programmed to switch from low-level hourly or daily sampling to more frequent sampling when one pod or a group of pods senses a potential hydrologic event, such as the start of an algal bloom or a change in stream stage in response to a storm event. This embedded computational intelligence is essential for applications in remote locations where researchers need to minimize human field effort, minimize energy requirements, and maximize sensor lifetime.

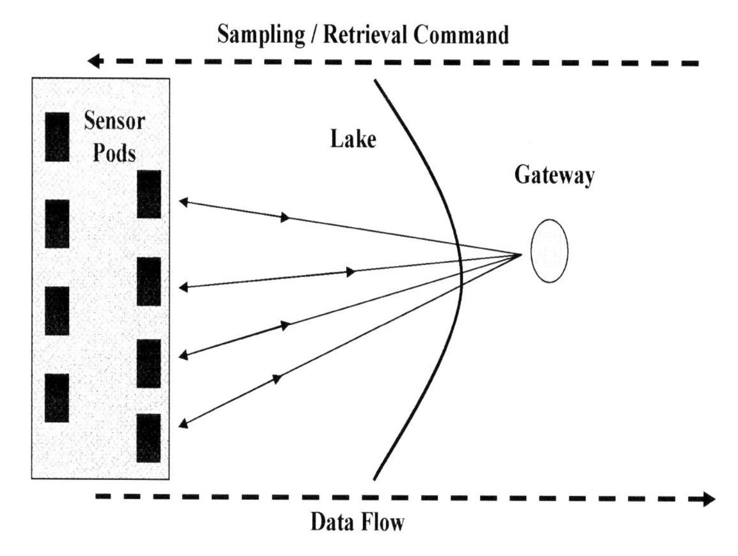

FIGURE 2-1 A schematic example of an embedded sensor network (ESN) in a lake environment. The shaded area indicates flexible networking connections between the pods with each other and the gateway. SOURCE: Seders et al. (2007). © 2007 by Mary Ann Liebert, Inc, publishers.

TABLE 2-1 Embedded Sensor Network (ESN)[a]

Embedded	Networked	Sensing
Embed numerous distributed devices to monitor and interact with physical world	Network devices to coordinate and perform higher-level tasks	Tightly coupled to physical world
Control system with, small form factor, wireless pods	Exploit collaborative sensing take action	Exploit spatially and temporally; dense, in-situ, sensing and actuation

[a] Or ENS or sensor web.
SOURCE: Adapted from http://research.cens.ucla.edu and http://sensorwaresystems .com.

In a typical ESN, each pod is equipped not only with sensors but also with a radio transceiver, a small microcontroller that communicates with the pods, and an energy source—usually a battery and where feasible a solar cell; although the sensor pods may also be hard-wired for some applications.

The sensing devices in an ESN are ideally small, robust, and inexpensive, so that they can be produced and deployed in large numbers. In order to be field deployable at low cost, device energy requirements, memory, computational speed, and bandwidth (due at least in part to FCC regulations) are typically constrained.

A key desirable feature of ESNs is that they have the ability to self-organize in order to cope with changes. Data from the sensors are usually aggregated and analyzed, either by a computer within the network or outside it. That is, the pods in an ad-hoc wireless sensor network are effectively self-organizing and hence do not require detailed, pre-programmed knowledge of network topology in order to function. Such networks will also be robust to a certain degree of network modification and will continue to function as new nodes join the network or existing nodes fail or move to new physical locations. The ESN, with its unique global information-sharing protocol, forms a sophisticated sensing tapestry that can be draped over an environment. This approach allows for various complex behaviors and operations, such as real-time identification of anomalous or unexpected events, mapping vector fields from measured scalar values and interpreting them locally, and single-pod detection of critical events, which then triggers changes in the global behavior of the sensor network (Delin et al., 2005). Note that a pod in an ESN is merely a physical platform for a sensor and thus can be orbital or terrestrial, fixed or mobile, and often has real-time accessibility via the Internet. Pod-to-pod communication is both omni-directional and bi-directional where each pod sends out collected data to other pods in the network. As a result, on-the-fly data fusion, such as false positive identification and plume tracking, can occur within the ESN, and the system subsequently reacts as a coordinated, collective whole to the incoming data stream. An example of how an ESN can be used for real-time management and control of combined sewage outflow is provided in Box 2-1.

It should be noted that ESN accomplishments have been modest to date and that their potential significance is unresolved. However, their upside potential is so high that it will be important to attempt to resolve the many challenges that will present themselves for field deployment.

In summary, an ESN is a network of small, sensor nodes or pods communicating among themselves using radio communication, and deployed in large scale (from tens to thousands, creating a sensor "web" at high density) to sense the physical, chemical, and or biological world. Unique characteristics of an ESN are as listed below (Maurice and Harmon, 2007, and other manuscripts in Environmental Engineering Science special edition, March 2007). This list describes the optimum or ideal ESN that the scientific community can strive for, but it is

BOX 2-1
An Embedded Sensor Network for Control of Combined Sewage Overflow Events

Many wastewater systems in the United States and abroad contain at least some components that were built to accept both sanitary wastewater and stormwater runoff, in order to minimize original construction costs. The problem with such systems is that during times of high flow, following storms or rapid snowmelt, the combined flow can become too great for sewage treatment facilities, and is then diverted directly into lakes and rivers, resulting in a combined sewage overflow (CSO) event. Following passage of the Clean Water Act in the early 1970s, municipalities were faced with a mandate to prevent such CSO events for the sake of the ecosystem and human health. Although many cities have invested millions to billions of dollars in updating their sewage systems, each year in the United States, CSO events still result in the release of hundreds of billions of gallons of untreated wastewater into lakes and rivers.

In order to help municipalities to minimize CSO events, low-cost, embedded sensor networks are being developed specifically to address the CSO problem. Recently, Ruggaber et al. (2007) developed and deployed an embedded sensor network in South Bend, Indiana (USA). This system was designed to decrease the frequency and severity of CSO events by maximizing the existing storage capacity already present in the city's combined sewer system. The embedded sensor network uses data gathered from a distributed network of sensors, many of which use wireless technologies, to provide decentralized, distributed, real-time control of the combined sewer system's storage capacity using automated valves called Smart Valves (Ruggaber et al., 2007). By decentralizing the decisionmaking and control, each section of the sewage system is able to maximize its efficiency, providing a finer control mesh that minimizes flooding. The embedded sensor network controls the storage of stormwater runoff in a large retention basin using level data from sensors within the basin and at the combined sewage system outfall, which is several miles away. Before the embedded sensor network was in place, the basin often was ineffective, resulting in CSO events following even relatively small storms. After emplacement of the network, the basin has been able to store all of the water that enters during almost all storm events, preventing a CSO event into the local St. Joseph River. Once the threat of a CSO event decreases, the sensor network system automatically releases the water stored in the combined sewage system to prepare for the next storm.

The real-time measurements within the combined sewage system can also be used to convert the existing CSO planning models into real-time control and operations models; to improve land management; to develop a real-time CSO public notification plan; to expedite maintenance; and to revise and improve the city's overall CSO strategy (Ruggaber et al., 2007). This is thus an example of how an embedded sensor network can be integrated with municipal operations and public services to provide substantial environmental benefits at relatively low cost and with much faster and easier deployment than traditional technologies. Ultimately, individual municipal embedded sensor networks can be combined into a national network to provide protection to streams and lakes on a regional to national scale.

likely that not all qualities will be attainable for all applications, given the many challenges of real-world environmental sensing.

- Small-scale sensor nodes
- Limited power requirements that they can harvest or store
- Ability to use in harsh environmental conditions
- Self-recognition of node failures
- Adaptive monitoring, actuated sample collection
- Potential real-time control of a local environment
- Optimal management of power consumption related to sensor excitation and wireless data transmission
- Potential mobility of nodes
- Dynamic network topology
- Self-recognition of communication failures
- Heterogeneity of nodes
- Large-scale deployment
- Unattended operation
- Web-based data and interaction (e.g., visualization)

The Need for New Sensor Probe Technologies

Fortunately, in concert with development of ESN technologies, sensors themselves are becoming increasingly smaller, more robust, and less expensive. In particular, physical sensors such as those that measure air and water temperature, water pressure, radiation, relative humidity, and wind speed and direction have evolved over decades and are now mass-produced and routinely packaged together in small instruments along with power and communication devices (Delin, 2002; Szewczyk et al., 2004; Vernon et al., 2003; Woodhouse and Hansen, 2003; Yao et al., 2003). These small and often inexpensive instruments often can be left in the field unattended for longer periods of time than traditional instruments, although problems of fouling, drift, etc. still need to be addressed. However, current physical sensor technology does not provide three-dimensional parametric physical information for air, water, soil, and groundwater over spatial scales ranging from the micro-scale (e.g., pore volume) to the kilometer-scale, and integrating a web of sensors measuring at different scales (e.g., web pods with kilometer-scale satellite measurements) is poorly developed.

In contrast to hydrometeorological variables, sensor development for many important chemical and biological measurements is relatively immature (Estrin et al., 2003; NSF, 2005). For example, sensors that measure nutrients in soil and water (e.g., phosphorus, nitrate) remain relatively expensive and are subject to rapid fouling or degradation. Chemical sensors are needed to measure a wide range of elements and molecules for inorganic, organic, and biochemical molecules in all environmental media (atmosphere, soils, sediments, groundwater, and fresh and marine waters). In particular, reliable means for detecting toxins and determining the presence and amount of nitrogen and phosphorus forms are critically important. Biological sensor technologies in general are the least mature, but investments such as those described in the section below should result in tremendous scientific advances. Biological sensors can provide key information on the function and structure/composition of biologically influenced ecosystems in real time (NSF, 2005). Development of a wide range of field-robust chemical and biological sensors is one of the greatest challenges facing widespread deployment of sensor networks in the hydrologic sciences.

The characteristics needed by sensors within ESNs provide challenging design criteria: They need to be cheap enough to be widely disseminated, small, reliable, and robust. Specifically, they should have low power consumption (and/or use of renewable energy); be robust to temperature fluctuations; have low maintenance requirements; remain free of measure drift over the lifetime of deployment (or be able to self-calibrate or be capable of remote calibration); be resistant to deteriorating accuracy of measurements attributable to organic interaction with sensors (i.e., biofouling); and have environmentally benign components and operation. The sensor network itself has many requirements; it needs algorithms to help detect individual faulty sensors and to flag data accordingly, flexibility of platform location, intelligent and synergistic operation, methods to maintain data and physical security, and an ability to query sensors and possibly recalibrate them remotely. Thus, users have a long list of desirable sensor features, and such add-ons will tend to drive costs up.

NEW AND EMERGING BIOGEOCHEMICAL SENSOR APPROACHES AND TECHNOLOGIES

Over the past decades, a variety of new approaches have been developed for understanding and quantifying hydrobiogeochemical processes in the field. Some of these approaches have been well established and widely adopted; others are just emerging and may or may not live up to their initial promise to expand our observational abilities.

Biogeochemical field sampling began as a labor-intensive activity, requiring researchers to go to individual sites in the field and collect samples for field and laboratory experiments. In addition to being labor intensive, this sampling tended to be biased towards daytime sampling in good weather, that is, the summer field season. The development of automatic sampling devices such as the ISCO sampler, which could collect water samples over time in stored bottles for later retrieval, has proved helpful in expanding many studies, but this has numerous limitations, especially for parameters that need to be measured immediately. Datasonde devices can be used for sampling surface and groundwaters and can be equipped with a variety of probes from temperature and dissolved oxygen to chlorophyll and nitrate. Although they have the ability to collect data continuously, they are expensive to equip and operate and the probes are subject to drift and fouling, which limit their usefulness. Nevertheless, together with automatic samplers, they have led to significant advances in hydrobiogeochemical understanding. A variety of sensors and field analytical devices relying on optical, ultraviolet, and infrared measurements have become available (McDonagh et al., 2008); connected to a range of sampling devices, they can be used to make multiple measurements, and can be integrated into projects depending upon cost and robustness requirements (e.g., temperature fluctuations).

Reviewing all of the new biogeochemical sensor technologies would be well beyond the scope of this report. Hence, in the sections below, the focus is on several examples of emerging technologies that may prove useful in hydrobiogeochemistry. First, a series of microsensor applications is presented followed by a discussion of molecular- to nanoscale sensors. Then applications of in-situ microcosms are introduced, followed by robotic samplers for biogeochemical monitoring. Later in this report (Chapter 4), some potential applications of the more developed of these technologies and approaches to existing hydrologic observatories are identified. As global observation systems are implemented, it will be essential to integrate the wide variety of different sensing approaches, in order to take advantage of the increased scientific capabilities and understanding.

Microsensors

Microsensors are miniaturized devices, micrometer to millimeter in size. They measure physical and chemical quantities such as pressure, acceleration, temperature, speed, and chemical or gas concentration. Typically they convert such physical or chemical quantities into an electrical signal, which, when calibrated, can be used as a proxy for the quantity itself. Below, three examples of microsensor applications to environmental problems are presented: (1) oxygen and pH sensors applied to acid-mine drainage, (2) nitrate sensors for biogeochemical

chemical applications, and (3) "lab-on-a-chip" designs for detecting pathogens in wastewater.

Microsensor Arrays for Biofilm Activity Associated with Acid-Mine Drainage

Mining activities expose rock surfaces to contact with air and water. Any sulfides present in such mine tailings, waste rock piles, or underground cavities can be oxidized under these conditions. Such oxidation reactions can lead to highly acidic conditions and elevated concentrations of dissolved metals such as iron, manganese, lead, cadmium, mercury, zinc, and copper. This highly acidic, metal-rich water can pollute streams and rivers and is a major environmental problem throughout the world, from the coal regions of Pennsylvania to the silver mines of Bolivia. Biogeochemical processes play an important role in controlling metal dynamics within acid-mine impacted streams. The reaction rates of acid-mine drainage reactions are accelerated many orders of magnitude by metal- (e.g., iron) and sulfur-oxidizing bacteria. As in other bacterial processes, bacteria adhere to surfaces in aqueous environments and excrete a slimy substance or biofilm that attach them to mineral surfaces. Recently, Haack and Warren (2003) used microelectrodes to probe in situ the dissolved oxygen and pH profiles within biofilms in an acid-mine drainage stream associated with a nickel mine in Ontario, Canada (Figure 2-2). Diel profiling with the microelectrodes demonstrated that biofilm oxygen and pH gradients varied both spatially and temporally, demonstrating that the biofilms are highly dynamic biogeochemical environments. By combining this in-situ microelectrode field analysis with sampling of metals and microorganisms, the authors demonstrated that hydrous metal oxide (HMO) minerals within the biofilm exert important influences on the concentrations of metals such as Ni, Co, and Cr, and that the HMOs themselves are effected by seasonal and diel processes. Combination of such high-resolution mechanistic information with more traditional field sampling and remote sensing can help to understand better the physical, chemical, and biological behavior of the acid-mine drainage sources, and to develop more-focused monitoring and remediation strategies.

Some of the remote sensing methods that could be combined with these new measurements include airborne thermal infrared and the airborne visible and infrared imaging spectrometer (AVIRIS). As the acid-mine drainage reactions are exothermic and generate warm water, airborne thermal infrared (TIR) can be applied to identify hidden sources of acid-mine drainage (Sams and Veloski, 2003) discharging into streams. The content of alkalinity-generating rocks within watersheds affected by acid-mine drainage determines the capacity of the streams to neutralize the acid. It is possible to identify and determine the areas covered by carbonate minerals such as calcite within the watershed using AVIRIS (Dalton et al., 2004).

FIGURE 2-2 A microelectrode array probing pH and dissolved oxygen gradients in a biofilm associated with acid-mine drainage, Ontario, Canada. SOURCE: Reprinted, with permission, from L. Warren, McMaster University, Canada.

Scaleable Nitrate Microsensors in the Form of a Plant Root

Microfabrication processes are enabling the creation of mass producible sensors in form factors more attuned to the environmental media in which they are deployed. For example, flexible, miniature, and inexpensive nitrate sensors are being fabricated by electropolymerizing pyrrole onto carbon fiber substrates, using nitrate as a dopant (Bendikov et al., 2005). The resulting sensor (length 1 cm or less; diameter 7 to 10 um; Figure 2-3) is a size and shape that is ideal for deployment in the pore space of soils and sediments.

In addition, these highly selective electrodes require no excitation voltage and can quantify nitrate concentrations to just below 10^{-4} M (3 ppm). In-situ tests have revealed that prototypical microsensors of this type are short-lived in the environment, losing their sensitivity in a span of hours to days (depending on conditions), but can be reconditioned and redeployed within 12 hours. Research is underway on modified fabrication techniques and materials (Hatchett and Josowicz, 2008) as well as microfluidic strategies for sample pretreatment and may soon produce more robust microsensors for long-term deployments.

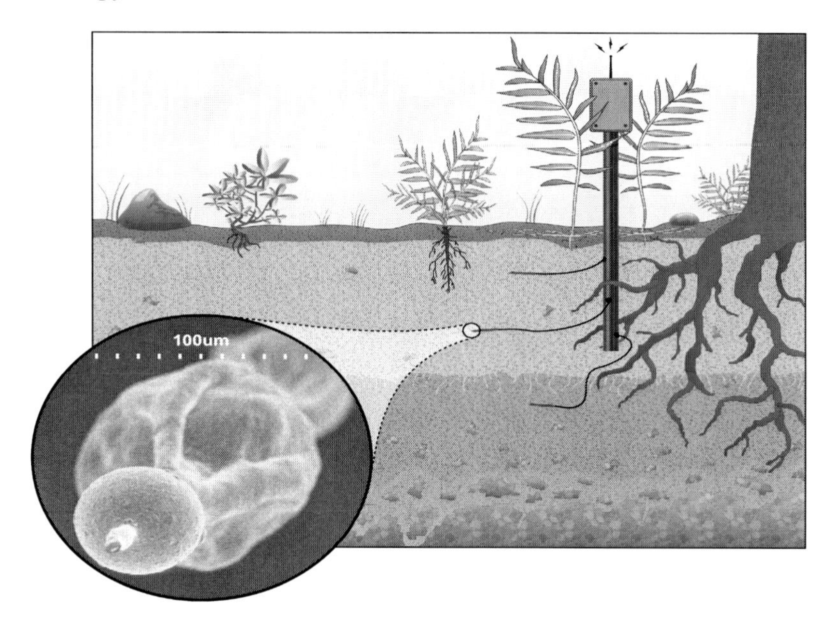

FIGURE 2-3 Illustration of an *in-situ* nitrate microsensor. SOURCE: Reprinted, with permission, from Jason Fisher, University of California, Merced.

"Lab-on-A-Chip" for Detecting Pathogens in Wastewater and Real-Time Process Control

Detection of pathogens in water samples is an important concern in environmental engineering, from protecting the water supply to managing wastewater. Ideally, one would like to have a small, cheap, efficient, and reliable detection system that could provide "real-time" data for management and control; i.e., a "lab on a chip" that could be tied to an ESN. The need to culture samples for analysis has made the identification of specific pathogens cumbersome and time consuming. Over the past decade, the development of molecular methods for evaluating microbial diversity and for detecting specific microorganisms without the need for cultivation has provided new hope for fast, easy, accurate, real-time monitoring.

Gilbride et al. (2006) review and evaluate recent progress in molecular microbiologic approaches, including discussion of microarray assays and of the state-of-the-art in the microfluidics and optoelectronics needed to realize sensor systems. They point out that one of the greatest impediments to development of detection systems for pathogens in wastewater and other environmental samples is the presence of numerous other compounds (organics, metals, etc.) that can cause interferences throughout the process, from sample collection to DNA purification steps to amplification to signal detection and interpretation. Nonetheless, the authors conclude, "Albeit the limitations confronting lab-on-chip designs, they still hold much promise for real-time monitoring of wastewater effluent since rapid detection of pathogens from wastewater could reduce the risk of release into the environment." It is important for the environmental community to keep track of advances occurring in the biomedical community, which has been intensely focused on development of microbial sensor systems.

Molecular- to Nanoscale Environmental Sensors

In recent years, the measurement of chemical parameters has turned to new technologies that consider the special properties present in structures of matter with at least one of the dimensions of the order of less than 100 nm or 10^{-7} m (e.g., Riu et al., 2006; Jianrong et al., 2004). The development of nanoscience and nanotechnology has opened the door for a new generation of "nanosensors" for various applications, including biomedical and environmental. For example, adsorption of a chemical species on the surface of bulk material is different from adsorption on the surface of a very thin film. The electrical properties of the film change. This change can be used to estimate the concentration of the adsorbed material, leading to development of new film-based sensors. Some other structures used as nanosensors include nanoparticles, nanotubes, nanowires, embedded nanostructures, and porous nanosensors (Riu et al., 2006). Application of such techniques to aqueous samples of environmental interest will depend upon development of prefiltration and potentially in some cases pre-concentration methods, and although the films tend to be very sensitive to individual chemical species, the potential for interferences from other components in an environmental sample will need to be addressed. Development of nanosensors for airborne materials, including contaminants, is in some cases less challenging and more progressed.

Nanosensor Types and How They Work

Due to their size, nanoparticles behave as quantum dots with wavelengths predicted by the wave-corpuscle duality principle. Particles with different sizes produce quantum dots with different wavelengths or energy levels. Nanoparticles (NPs) bound to biological molecules have been used in biosensors to detect and amplify an analyte. Carbon nanotubes (CNTs) are concentric cylinders a few nanometers in diameter. Length can be up to hundred of micrometers. They can be single-walled or multi-walled nanotubes. The electrical conductivity of nanotubes changes when molecules of the analyte approach the walls of the nanotube. For example, the electrical conductivity of a single-walled nanotube changes in the presence of oxygen (Collins et al., 2000) as well as in the presence of other gases. Carbon nanotubes and nanoparticles can be functionalized with molecules to interact with specific analytes making them feasible for many possible applications.

Nanowires have been used to detect several chemical species, such as NO_2 (Zhang et al., 2004). Boron-doped silicon nanowires (SiNWs) have been used to detect Ca^{2+} and pH (Cui et al., 2001). For pH, the SiNWs were functionalized with amine and oxide. The change of conductance with pH was linear and can be explained as a change in surface charge during protonation and deprotonation. Embedded nanostructures are nanosized microstructures within bulk solid materials, usually the microstructures are nanoparticles (Riu et al., 2006). They are like assemblies of nanoelectrodes. Nanosensors that use this method include a sensor for SO_2.

Nanoporous polymer membranes have been used to construct low-cost humidity nanosensors (Yang et al., 2006). They can be fabricated with inexpensive materials such as polycarbonate, cellulose acetate, and nylon. Nanopore humidity sensors can detect the changes in resistance and/or capacitance due to water adsorption inside the nanopore cavities, which have a very large surface area. The porous membrane can sense the humidity because of ionic conduction. The adsorbed layer of water at the pore surface reduces the total sensor impedance as the ionic conductivity and the capacitance increase. Nanosensors to determine H^+, Ca^{2+}, K^+, Na^+, Mg^{2+}, Zn^{2+}, Cu^{2+}, and Cl^-, and other species have been constructed and encapsulated by biologically localized embedding (PEBBLEs; Buck et al., 2004). PEBBLEs are sub-micron-sized optical sensors designed to monitor an analyte concentration with minimum invasion and interference.

The development of nanosensors has occurred preferentially in the medical and biochemical field because of the need for real-time measurements of chemical and physical parameters with high resolution and negligible perturbation of the sample. These requirements also exist in environmental monitoring. One of the reasons for the preferential development in medicine and biochemistry is the ability to commercialize the sensor technologies for a large and potentially lucrative market. The environmental community needs to keep careful track of the biomedical community's sensor developments and to seek out opportunities to expand nanosensor use to samples of environmental interest. Because the environmental community has significant strengths in areas such as surface chemistry and dealing with complex systems, it can offer the chemical and electrical engineering communities' assistance with addressing issues such as biofouling, interferences, and surficial properties. The hydrologic community can work to offer expertise on the unique properties of water at surfaces and in nanopores/spaces.

Even when the different nanosensors are still in the development stage, they should be pursued for environmental applications because they could offer several advantages with respect to conventional probes: (1) they should be smaller and more portable, (2) they should be cheaper, (3) when properly packaged, they could be more robust and keep better and for longer time calibration than macrosensors, and (4) they have low energy needs and therefore could be easily incorporated into ESNs run off of batteries. Funding should be available to develop nanosensors to their full capacity. The challenge will be to make the novel sensors robust enough for field deployment and to calibrate and verify them against existing macroscale and laboratory analysis, where applicable.

An Example of Bacterial Surface Force Measurements

One example of the use of nanoscale technologies for environmental sensing is the use of the atomic force microscope (AFM) to conduct force measurements between the surfaces of bacteria and other surfaces of biogeochemical interest, such as minerals, organic films, and other bacteria (Figure 2-4). The AFM is formed by a microscale cantilever with a sharp tip or probe at its end. This tip is used to scan the specimen surface. When the tip is brought into proximity of a sample surface, forces between the tip and the sample deflect the cantilever. The deflection is measured using a laser spot reflected from the top of the cantilever and detected by an array of photodiodes. Microcantilevers are discussed in detail in Goeders et al. (2008).

For example, Lower et al. (2001) used the AFM to measure the forces between a dissimilatory metal-reducing bacterium, *Shewanella oneidensis* and the Fe oxyhydroxide mineral goethite. Force measurements were made with sub-nano-newton resolution in real time using living cells in aerobic and anaerobic solutions, as a function of the distance between the cell and the mineral surface, in nanometers. In another example, Lower et al. (2005) used AFM to measure the forces between a protein on the surface of living *Escherichia coli* bacterium and a solid substrate, in situ. Understanding the forces between environmental surfaces can help in developing models for a host of environmental processes ranging from how bacteria attach to minerals and form biofilms to the factors controlling retardation of pollutants in subsurface environments.

At the current time, the AFM is primarily a laboratory instrument, although AFMs can be taken into the field for use in mobile labs. AFMs have been manufactured to fit in the palm of one's hand or in a pocket, and to require very little power to operate. Hence, AFMs are even being considered for inclusion in manned missions to Mars (http://www. nasa.gov/missions/science/afmicroscope.html). Application of these and other nanoscale sensors to aquatic and terrestrial field sites will be challenging given concerns about sensor operation and lifetime time (e.g., Buffle and Horvai, 2000), including biofouling, calibration drift, need for prefiltration, interferences, and other factors.

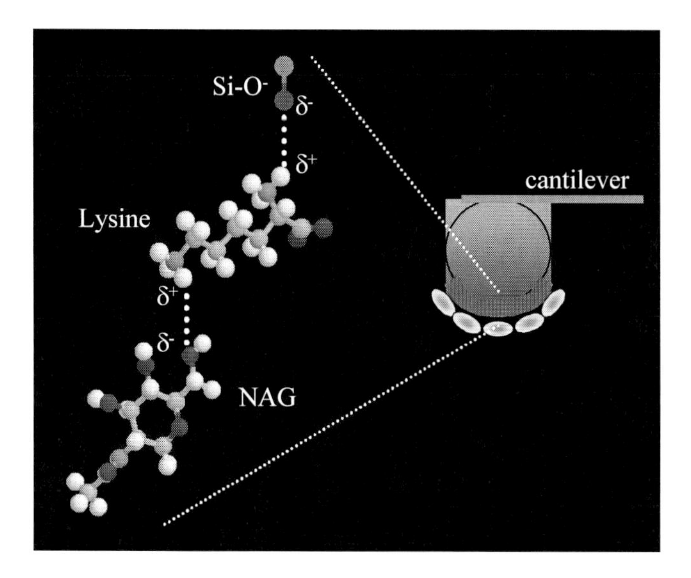

FIGURE 2-4 An AFM cantilever coated with bacteria attached to the probe using an organic "glue" for use in nanoscale force measurements. SOURCE: Reprinted, with permission, from Steven Lower, Ohio State University.

In-Situ Microcosms: Combining Field Experiments with Microbiologic and Surface Science Techniques

Although ESNs with real-time data acquisition and transmission promise to revolutionize biogeochemical research in the hydrologic sciences, many biogeochemical processes, such as mineral weathering, occur slowly and can benefit from other in-situ measurement techniques. The in-situ microcosm approach in biogeochemistry, wherein mineral surfaces are placed in "holey" containers in the field and excavated months to years afterwards for analysis by sophisticated surface science and microbiologic techniques, has proven highly useful for understanding mineral weathering rates and mechanisms (e.g., Bennett et al., 1996; Nugent et al., 1998; Maurice et al., 2002). Fundamental hydrobiogeochemical questions raised by more traditional sampling approaches can be explored mechanistically by use of these in-situ reactors that act as sampling devices at timescales appropriate for the processes in question, resulting in data integrated over an appropriate timescale.

Nugent et al. (1998) used an in-situ microcosm approach to investigate rates and mechanisms whereby feldspars, which are the most abundant minerals at the

Earth's surface, weather in an acidic soil. This research was conducted to address why many previous experiments had tended to suggest different rates and mechanisms of feldspar weathering in the laboratory versus the field. The albite feldspar surfaces were buried for up to several years, and analyzed by auger electron spectroscopy, X-ray photoelectron spectroscopy (XPS), and secondary ion mass spectrometry for surface composition and depth profiling, and by scanning electron microscopy and atomic force microscopy for particle micromorphology and microtopography. The results suggested that previously observed apparent differences in field versus laboratory rates of weathering could result at least in part by formation of a thick natural mineral coating on the feldspar surfaces that could not be detected by more conventional methods.

In another study, Maurice et al. (2002) used in-situ microcosms as an independent method to determine whether mineral weathering rates in the hyporheic zone (i.e., the wetted sediments surrounding a stream; Figure 2-5) of an Antarctic Dry Valley stream were as fast as indicated by stream chemistry measurements; the fast weathering rates appeared to be at odds with the fact that mineral weathering rates are thought to be slow at low temperature and in the absence of vascular plants. Mica chips were buried in in-situ microcosms in the hyporheic zone of Green Creek, a McMurdo Dry Valley stream for 39 days of the short Antarctic "summer".

AFMS and XPS revealed that the mica surfaces were quickly colonized by bacteria (Figure 2-6) and that small etch pits formed indicating a relatively fast weathering rate. Analysis of soil bacteria indicated a rich microbial community including nitrate-reducing bacteria. Overall, this study indicated the importance of combining different approaches, from traditional hydrologic and water chemistry analysis to microbiologic analysis, to application of surface science techniques, to help understand rates and mechanisms of biogeochemical processes.

The in-situ microcosm approach is being further expanded to include flow-through systems for investigating stream hydrobiogeochemical processes involving microorganisms. Microbial communities are often highly complex, and their biogeochemical behaviors are best studied in the field. This is because microbial communities and their biogeochemical behaviors are essentially impossible to maintain in the laboratory. Professor Gill Geesey and graduate student Eric Boyd at Montana State University have developed in-situ continuous-flow reactors (Figure 2-6) that can be emplaced in field sites and allowed to incubate for extended periods of time. The researchers have used these in-situ continuous flow reactors, coupled with geochemical and molecular microbial analysis, to investigate directly carbon fixation by sulfur-associated microbial populations in an acid geothermal spring in Yellowstone National Park (G. Geesey, Montana State University, pers. commun., October 2006). By using such an in-situ method,

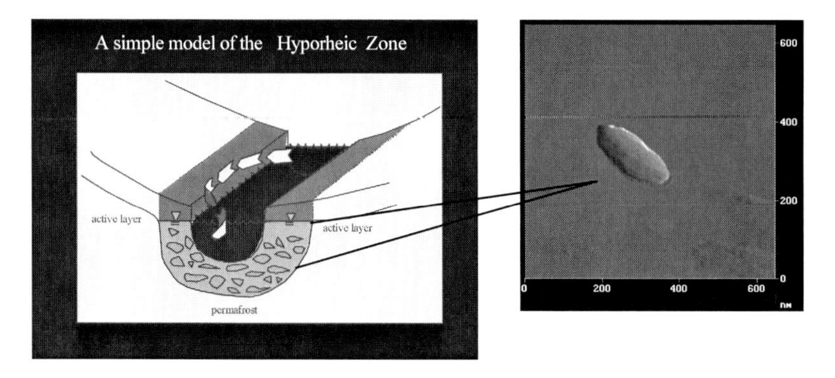

FIGURE 2-5 Left: A schematic of the hyporheic zone of an Antarctic Dry Valley stream. The lower boundary of the hyporheic zone is well-defined by the permafrost. SOURCE: Reprinted, with permission, from Diane McKnight, University of Colorado. Right: An atomic force microscopy image of a bacterium on the surface of a mica particle buried in the hyporheic zone. SOURCE: Reprinted, with permission, from Maurice et al. (2002). © 2002 by Elsevier.

FIGURE 2-6 In-situ continuous flow reactors for microbiologic community and activity sampling. SOURCE: Reprinted, with permission, from E. Boyd and G. Geesey, Montana State University.

the microbial community composition and physiological activity can be determined in the field, thus helping to ensure that the results are indicative of actual field processes.

Arsenic in drinking water can pose a serious human health risk; arsenic has been linked to a variety of different types of cancer, to serious skin conditions, and to nervous system damage. Professor Janet Hering at the Swiss Federal Institute of Aquatic Science and Technology (Eawag) has been leading a study to determine the controls on arsenic mobility in the Los Angeles Aqueduct system (e.g., Kneebone and Hering, 2000). An important component of this research is determining the concentration and mobility of arsenic in pore waters of surficial sediments in the aqueduct management system. This can be challenging because arsenic speciation can be disturbed by traditional sampling methods. Her group has used a gel probe sampler composed of polyacrylamide gel slabs held in a Plexiglas ladder, which is placed in the sediment and allowed to equilibrate with the pore water for several hours. The sampler is then removed and the pore water is analyzed in the laboratory. Recently, Hering and her associates have modified the sampler by doping the gel slabs with Fe(III) oxyhydroxide to provide further information on arsenic adsorption to reactive mineral surfaces (Root et al., 2005). Like the in-situ microcosms describe above, this novel sampler provides data from a given field site at a sampling time/period appropriate for the hydrobiogeochemical processes of interest and in a manner that merges field and laboratory methods to provide high-quality analysis of in-situ field conditions.

Robotic Samplers for Aquatic and Terrestrial Biogeochemical Monitoring

As described above, ESNs are expanding opportunities in high-resolution, real-time environmental monitoring and control. In many instances, such as for investigating spatially and temporally variable biogeochemical processes in lakes and in the marine environment, it may be beneficial to have mobile sensor platforms or pods integrated into the sensor network. Ongoing research by G. Sukhatme and colleagues (Sukhatme et al., 2007; Dhariwal et al., 2006) focuses on development and test deployment of a mobile robotic boat operated in conjunction with a static network of buoy nodes (Figure 2-7). The static buoys provide "low-resolution spatial sampling with high temporal resolution while a mobile robotic boat provides high resolution spatial sampling with low temporal resolution" (Dhariwal et al. 2006). The robotic boat performs autonomous global positioning system (GPS) way-point navigation between the static nodes, collecting biogeochemical data along the pathways. Sensors used in test deployment include thermistors, along with fluorometers to measure the concentration of chlorophyll-a, which is a key indicator of the presence and activity of

FIGURE 2-7 A robotic boat for microbiologic sampling. SOURCE: Reprinted, with permission, from Dhariwal et al. (2006). © 2006 by IEEE.

certain photosynthetic microorganisms. The combination of static buoys and a robotic mobile sensor unit represents an intriguing advancement in real-time measurement of environmental parameters in aquatic environments and furthers the possibility for sensor-actuated sampling to provide excellent coverage both spatially and temporally.

In another application, Harmon et al. (2007) have developed a rapidly deployable networked infomechanical system (NIMS RD) technology for measuring hydraulic and chemical parameters across stream channels. The system consists of two supporting towers and a suspension cable that delivers both power and Internet connectivity to control and actuate a tram-line NIMS unit, which raises and lowers a sensor pod either according to a pre-programmed route or a data-actuated adaptive scan. This tram-like robotic device has been test deployed in a small urban stream in Southern California and the San Joaquin River in Central California, and has been used for both nutrient (nitrate and ammonium sensors) and salinity sampling, in addition to temperature and pH. Because the system is rapidly deployable, it could potentially be moved around a geographical area to monitor sudden events such as combined sewage overflow to rivers.

Hamilton et al. (2007) developed a NIMS tram system as part of a terrestrial ecological observatory at the University of California James San Jacinto Mountains Reserve. The NIMS tram system (Figure 2-8) was designed to integrate with fixed sensors (soil sensors for CO_2, NO_3, H_2O, temperature, soil moisture content along with soil microrhizotron) to help support modeling of soil biological activity and CO_2 flux. The mobile robotic NIMS unit is mounted to a cable and uses an IR thermal scanner system to map the soil surface temperature, along with above-ground profiling of wind, incident solar radiation, and relative humidity above each of the fixed soil sensor nodes. In addition, a three-dimensional laser scanner is used to build a model of the tree and shrub canopies to collect data for modeling of solar interception and to produce volumetric biomass measurements. This network of fixed and mobile sensors is providing continuous measurements of microclimate and soil data at a spatiotemporal resolution that our understanding of forest dynamics into global carbon models (Hamilton et al., 2007).

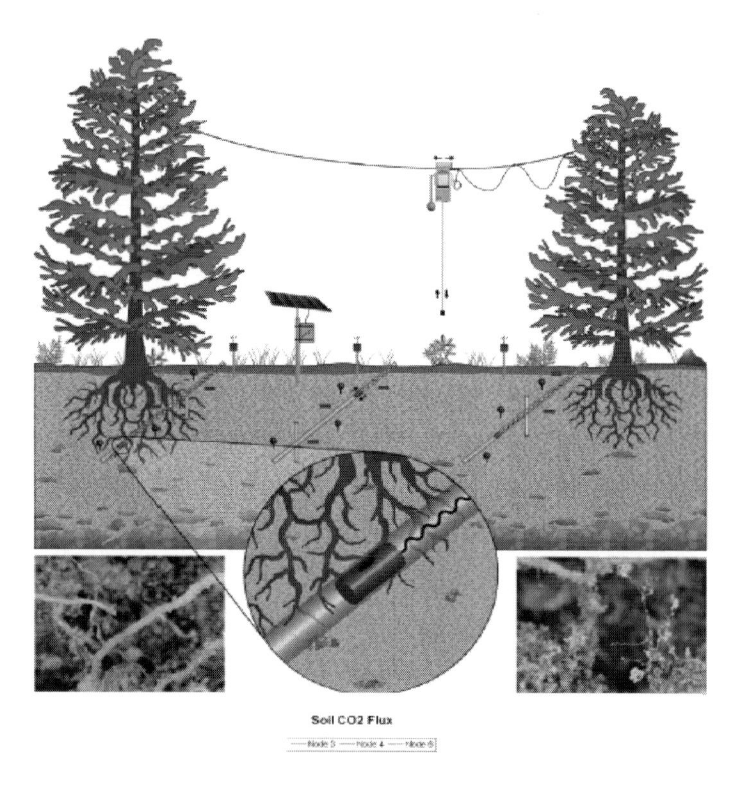

FIGURE 2-8 A NIMS tram for microclimate measurements, coupled with soil measurement techniques. SOURCE: Reprinted, with permission, from Hamilton et al. (2007). © 2007 by Mary Ann Liebert, Inc.

Interdisciplinary Challenges

As discussed above, all of the novel and emerging approaches described above share the common trait of requiring interdisciplinary cooperation and collaboration, not only within the scientific and engineering communities, but also sometimes with other communities, regulators and managers, and corporations. Within academia and many government research organizations, grants for research funding are obtained based on novel research ideas and significant new advances, with less funding available for continued development. Dissertations in electrical and chemical engineering can be built on development of a novel environmental sensor, but the field testing and implementation are ultimately the realm of the environmental scientist/engineer.

For novel sensors to become available at low cost and in abundance, they will need, in most cases, to be turned over from scientific researchers to the private sector, which means that they have to be commercially viable. For example, hydrobiogeochemical sensors that are useful not only for research but also for municipal and domestic water use increase the commercial viability and the probability that corporations will work harder to develop, test, and commercialize the sensors. In another example, computer scientists and engineers can write dissertations on optimizing codes for ESNs, including detecting faulty data, and new data management schemes and for novel computational methods. But, a dissertation only lasts a few years until the project is passed to the environmental scientists for long-term use and management, along with potential further code development.

Importantly, a significant market presence can encourage creation and acceptance of industry standards for sensor compatibility, communication, and sensor metadata. For instance, standardized approaches that automate capture and encoding of sensor metadata can facilitate the process whereby sensor data are ingested, quality assured, transformed, analyzed, and converted into publishable information products (Michener, 2006). Automatic metadata encoding should also enable scientists to track data provenance throughout data processing, analysis, and subsequent integration with other data products.

The hydrologic community has a good deal to offer electrical and chemical engineers, computer scientists, and other researchers working on sensors and sensor systems by providing new types of applications, experience with complex systems, detailed knowledge of water properties and flow, and specific expertise with a variety of Earth materials. Ultimately, in order for many of the emerging sensing approaches to succeed, it is important for agencies and foundations to facilitate collaboration between researchers in different fields, not only through short-term initiatives but also in the long term, and to assist scientists and engineers in building successful partnerships with corporations and communities.

AIRBORNE SENSORS

Airborne hydrologic sensing systems are potentially powerful, but infrequently used, elements of comprehensive hydrologic observing systems. While airborne systems are commonly used for high-resolution topography, land-use/land cover classification and change, and other purposes, they are infrequently used by hydrologists designing observing systems. This is because they tend to be expensive relative to in-situ sensors, the time required to develop a competent airborne system when one is not readily available is often long relative to the time available for the anticipated field project, and lack of access to competent airborne systems has limited development of the interpretive models needed to fully use airborne data.

However, there are several hydrologic applications that are common enough to have been taken up by the private sector. For example, high-resolution TIR imagery can be used for water temperature monitoring. Applications include characterizing point source thermal pollution, mapping aquatic habitat for threatened and endangered fish, and identifying groundwater discharge zones (including acid-mine drainage sites; see earlier section on "Microsensor Arrays for Biofilm Activity Associated with Acid-Mine Drainage"). For applications with high spatial- and spectral- resolution requirements, the Airborne Imaging Spectroradiometer for Applications (AISA) sensor system is commonly used. L-band passive microwave instruments also have shown their value in soil moisture mapping (see section on Spaceborne Sensors).

Interferometric synthetic aperture radar (InSAR), which measures elevation changes of the ground surface, has shown considerable promise for a number of water-related applications. These include timing, magnitudes, and patterns of seasonal ground movement in response to groundwater pumping and artificial recharge, estimating the elastic storage coefficient of an aquifer system (in conjunction with in-situ water table observations), and as a constraint for inverse modeling of regional groundwater flow (Alley et al., 2002).

The value and potential of airborne sensors are best understood by distinguishing among the distinct purposes for such systems. Airborne hydrologic sensing systems are developed to serve in one or more of three categories:

1. As an operational system to extend observations beyond the spatial domain of point measurements or to fill the gap between plot-scale observations and satellite-scale observations,
2. As a simulator of a satellite instrument to calibrate and/or validate data from a current or future satellite observing system, and
3. As a research instrument to support investigations of hydrologic processes where physical access to the spatial domain of a process is limited or where providing an adequate density of point measurements to characterize a process would be impractical.

Operational federal agencies, like NOAA and the USGS, or state agencies charged with water resources management, are likely to sponsor systems in Category 1. An airborne gamma ray system to measure Snow Water Equivalent (SWE) is an example of a Category (1) system. The National Weather Service routinely offers gamma flights to supplement the 1700 point measurements of SWE at snow pillows and transects of calibrated snow courses using airborne gamma ray sensors (e.g., Duval, 1977). However, data must be corrected for soil moisture. Gamma flight lines do not provide complete watershed coverage, but do provide much better spatial coverage than do the ground-based measurements. In "Achieving Predictive Capabilities in Arctic Land-Surface Hydrology" (Chapter 4), the role of airborne sensors as an integral part of an observing system is described. For example, airborne gamma ray, typically operated by contractors, is an operational element of regional water storage management systems for the northern prairie (Fritzsche and Burson, 1973).

Airborne data collection might also be used to help design terrestrial- or aquatic-based sensor networks or sensor webs. Where this is appropriate, they would be similarly useful in helping to oversee, calibrate, and integrate a sensor network once it has been established. Airborne systems can also be invaluable for scaling between sensor networks and satellite-based observations.

Until recently[1], the National Aeronautics and Space Administration (NASA) has had primary responsibility for developing satellite environmental sensing technologies. In that role, NASA produced several Category 2 airborne hydrologic sensors. Examples include the Electronically Scanned Thinned Array Radiometer (ESTAR; Levine et al., 1990), the Airborne C-band Microwave Radiometer (ACMR; Kim et al., 2000), and the Passive and Active L and S band (PALS; Wilson et al., 2001) microwave instrument. While none of these were ever to be exact simulators of future satellite sensors, each incorporates essential aspects of an anticipated satellite sensor. Because NASA focused primarily upon Category 2 objectives, developing global environmental observing technologies, and not Category 3 objectives, addressing research opportunities in hydrology, the airborne platforms chosen to carry NASA's sensors were either large, long-range aircraft capable of deployment to any spot on the globe, or aircraft capable of flying in the stratosphere above the 16 km altitude limit of most commercial aircraft. These NASA aircraft have been expensive to operate, difficult to operate safely at speeds below 90 m/s or at altitudes below 300 m above ground level, and unable to operate out of small airports or unimproved fields. Some consequences of this selection of platforms have been that the cost of using NASA's airborne hydrologic sensors have prohibited near-daily use in season-long field campaigns, or deployment to smaller campaigns in difficult locations, e.g., they have never been used in Arctic land-surface hydrology. Where they have been used, the difficulty of safely flying low and slow has

---

[1]In February 2006, NASA released a revised mission statement deleting "To understand and protect our home planet". The change is consistent with its recent focus upon the human exploration of the Moon and Mars.

meant that the data from these systems have often to satellite-scale gap in spatial-resolution of low spatial resolution sensors like microwave radiometers.

NASA's Category 2 airborne systems have been used in a Category 3 capacity, that is, to examine hydrologic processes that extend beyond the spatial domain of point measurements. While not ideal for this role, there are few other systems available for Category 3 purposes. As inaccessible as NASA airborne systems have been relative to the need, they have become even less accessible with NASA's change in focus away from the Earth sciences.

Category 3 airborne systems, those used in support of research, typically employ airborne platforms that are less expensive and have capabilities that are optimized for the objectives of the research. Examples of this include sensors on light commercial and homebuilt aircraft (e.g., Baldocchi et al., 1996), ultralight aircraft (e.g., Junkermann, 2000), helicopters (e.g., Hyyppa and Hallikainen, 1993), remotely controlled blimps (Inoue et al., 2000), and unpiloted airborne vehicles (UAVs) (e.g., Schoenung and Wegener, 1999). These systems contribute greatly to advancing our research by enabling observations at spatial and temporal scales that are otherwise inaccessible and because they are typically under the direct control of the science investigators. They are also challenging to develop and maintain. Airborne sensor systems are expensive relative to in-situ systems, their development typically requires several years before the data they produce are reliable, and effective interpretation of data from new airborne sensors often requires development programs that can be unique to the proposed application.

Table 2-2 offers an approximate comparison of operating costs. These costs do not reflect the time to develop the sensor, integrate it with the airborne platform, or successfully complete the test, "make-work", and system validation phase of development. Of these, the test, "make-work", and system validation phase is often underappreciated and under-funded. Care should be taken to reserve sufficient funds for this phase of development so that airborne systems do not arrive in the field before they are ready to produce reliable and interpretable data.

The difficulty and experience required to develop competent airborne systems favor establishing frequently reviewed centers having this competence. The airborne platforms often have to be modified with holes for sensors, with enhanced electrical systems that are free of electrical noise, and with command and data systems that guide surveys of pre-determined flight lines and recover observed and instrument performance data that can be co-located with position, attitude, and environmental data. The instrument itself often must be designed to be smaller, lighter, and/or consume less power than previous examples of similar instruments. They often need to function in a hostile environment that includes vibration, large excursions in temperature and pressure, noisy power sources, and typically constrained opportunities for calibration. Centers like the

TABLE 2-2 Examples of Airborne Platforms Capable of Carrying the Canonical Hydrology Sensing Instrument (for general comparative purposes only; cost estimates in 2007 dollars)

Class Aircraft	Source[a]	Example	Payload[b]	Approx. Cost/Flight Hr[c]	Daily Rate (Labor)[d]
Fixed-Wing, single piston engine	Commercial utility, new	Aviat Husky A-1B	300 lb/135 kg	$200	$500
	Homebuilt utility, new	Zenith CH 801	450 lb/200 kg	$200	$500
Fixed-wing – single turbo-prop engine	Commercial utility, used	Fairchild/ Pilatus Porter	1500 lb/680 kg	$1,500-$2,000	$1,000
Fixed-wing - twin turboprop engines	Commercial utility, used	DeHavilland Twin Otter	1500 lb/680 kg	$750	$1,500
Helicopter, single turbine engine	Commercial utility, used		1000 lb/450 kg	2,000	$1,500
UAV, single piston engine	Commercial utility, new	Predator	300 lb/135 kg	About $1 million/month (~$30,000-40,000/day)	
UAV, single piston engine	Commercial utility, new	Aerosonde	11 lb/5 kg	$1,000-$1,500	$4,000-$10,000
Fixed-wing, 4 turboprop en-gines	Commercial utility, used	Lockheed C-130	20,000 lb/ 9,100 kg	~$10,000	$2,000

[a]The choice of new or used aircraft is based upon the likelihood of finding one in suitable condition. The ratio of flight time to ground preparation time for aircraft used in research is typically very low relative to other uses of aircraft. A few hundred hours of flying time per year might be the norm. This level of utilization does not require a low-time airframe or engine, but the research application does require high reliability and safety, which translates to a requirement that the aircraft be in excellent condition. Single engine, piston utility, and homebuilt aircraft have a higher likelihood of fatigue or hidden problems than the other categories of aircraft. For these aircraft, project risk is reduced with new aircraft.

[b]Payload is payload listed for aircraft in a typical configuration less the weight of required crew and the weight of fuel in main fuel tanks.

[c]Approximate cost per operational hour. This does not include amortization of the platform or instrument, nor does it include the cost of the special support crew necessary to maintain and operate the aircraft during a field campaign where the aircraft is not at its normal home base. These costs are major cost drivers depending on the aircraft, and can range from a low of one person in the field who supports the instrument, does aircraft maintenance, and flies the plane, to as high as 20 personnel that are required to support the Predator. All other aircraft fall in between. A rule of thumb, except for UAV systems as the Predator, is the number of support personnel required to go into the field equals the number of engines on the aircraft. The flight hour costs posted in the above table cover primarily the direct operating costs of the aircraft. It is assumed that a member or members of the science team are qualified to pilot the fixed-wing, single-piston engine, aircraft such as the Aviat Husky or the Zenith CH 801. If the pilot is not qualified, then additional daily costs for the pilot would apply. Despite these costs, it is generally recommended to lease these services rather than purchase an aircraft.

[d]Assumes about $500/person, which includes salary and travel.

SOURCES: G. W. Postell and A. R. Guillory, Wallops Flight Facility, NASA; A. W. England, University of Michigan.

NOAA airborne facility at Boulder, Colorado, and a few university laboratories do provide high-quality airborne instruments. Maintaining these competencies should be a concern of the hydrologic community and the federal funding agencies.

Comprehensive descriptions of all potential airborne platforms would fill volumes. The span of platforms appropriate for hydrologic research can be greatly reduced by characterizing a reference sensor or set of reference sensors that a subset of airborne platforms might accommodate. One reference sensor might be an airborne SWE profiler that employs nadir-viewing, 1.4, 6.9, 19, and 37 GHz radiometers. Using the best of today's technologies, such a system might weigh 100 kg, consume 200 W, require external antennas with a combined aperture of 1 m^2 and a thickness of 10 cm, occupy a volume inside the aircraft of 30 cm^3, and cost less than $150/flight hour to operate not including the cost of the pilot, amortization of the system, or the cost of the radiometer system. It would fly safely during daylight hours at altitudes above ground level of between 30 m and 1 km, at speeds as slow as 30 m/s, and have the endurance to cover 500 km without refueling. Of the potential airborne platforms, only commercial and homebuilt light, fixed-wing aircraft and UAVs might have this performance envelope. Table 2-2 offers examples of suitable platforms along with examples of platforms that might be suitable for other reference sensors. The C-130, a four-engine turboprop aircraft used by NASA, is included in Table 2-2 for purposes of comparing operating costs. It is shown in Figure 2-9.

Homebuilt aircraft kits are listed, because they often possess characteristics that are of value in Category 3 research projects. Homebuilt aircraft kits tend to be small, relatively inexpensive to purchase, simple to fabricate, and inexpensive to maintain, because the builder is legally allowed to perform required maintenance. Because the builder is responsible for the structural integrity of the aircraft, they are relatively simple for a team with access to aircraft design talent, as would be the case with most engineering colleges, to add an instrument pod or to pierce the aircraft skin for instrument cables. These characteristics have been successfully used for research purposes (Oechel et al., 2000). If conservative designs using proven aircraft-quality engines from kit manufacturers having a history of, perhaps, hundreds of successful fabrications are chosen, there is little physical risk beyond the normal risk of operating light aircraft safely.

UAVs capture attention for their potential of avoiding the risk and expense of protecting pilots from challenging environments, e.g., altitudes above 16 km, and to enable use of smaller platforms that need not accommodate a pilot. UAVs are expensive to develop and can be expensive to operate if their support systems are complex. They have been developed primarily by the Department of Defense to avoid placing pilots in hostile battle environments, to enable observing systems that are small enough to be difficult to detect, and, where the

(a)

(b)

FIGURE 2-9 High-end and low-end airborne sensing. (a) A Lockheed C-130 airplane, extensively modified by NASA for use as a research platform and supporting a wide variety of onboard sensors, including multispectral scanners, radiometers, air sampling equipment, and aerial cameras. SOURCE: Available online (b) A de Havilland Canada Beaver configured for airborne geophysics in the 1970s. The primary instrument was a 1.4 GHz profiling radiometer. SOURCE: Photos courtesy of Gordon Johnson, U.S. Geological Survey.

objective is simply imaging, to enable the use of simple systems, like the Predator in Table 2-2, in a battlefield environment without complex support systems. If there were a UAV system that met the requirements of our reference sensor, its size and support requirements would approach those of a small, piloted aircraft, and, because of its newness, operation in populated areas would present safety issues for the air traffic control system and for people on the ground.

There are significant advantages in research effectiveness to attaching research aircraft to the facilities that build and operate the research instruments. This is a characteristic of the highly successful NOAA airborne operation in Boulder, Colorado. Universities often serve as operators of the simpler of these aircraft, but they typically lack sufficient volume of instrument building and community demand for field use of their instruments to justify dedicated aircraft at the more complex and expensive end of the platform spectrum. One comprehensive Category 3 research program, like the NOAA operation, and several university programs that focus upon new instrument technologies utilizing the less complex airborne platforms might serve as an optimum combination to support hydrologic research.

SPACEBORNE SENSORS

Satellite remote sensing plays a critical role in establishing a framework for integrating hydrologic and environmental observations over space and time. Spaceborne sensors offer an unparalleled, synoptic view of the landscape, providing a natural template for upscaling and extrapolating traditional hydrobiogeochemical measurements, in-situ sensor network, and airborne measurements. On the other hand, the direct terrestrial or aquatic and airborne measurements can serve to "ground-truth" satellite remotely sensed data and to provide additional data that cannot be remotely sensed, such as information on details of biochemical reactions in soils or microbial community analysis. Routine collection of satellite data allows for monitoring the dynamics of natural systems in space and time, while the space-based, global view enables interpretation in the context of larger-scale Earth system processes. Moreover, satellite platforms can serve as pods in embedded networks (i.e., as spatially distributed sensor platforms that wirelessly communicate with each other), providing real time information on the state of the observed system, or prediction of future states, that could trigger adaptive behavior of the sensor web. Satellite measurements are sometimes the only measurements available in locations that are remote or inaccessible for political reasons. As such, spaceborne measurements are a key element in data assimilation or any other viable scheme for integrating observations.

In order to realize the full potential of satellite remote sensing for enhancing integration strategies, its current limitations must be well understood in order to focus research priorities. Here three critical areas for research—rather than an exhaustive list—are described as examples of the significant challenges that lie ahead for remote sensing of hydrologic and related parameters. These areas include the impact of heterogeneity on enhanced retrieval algorithms, enhanced antenna engineering, and the role of geodetic satellites in hydrology. The focus here is limited to liquid water and snow. Although the dynamics and mass balances of glaciers and ice sheets are well monitored by satellite (e.g., Rignot and Thomas, 2002; Velicogna and Wahr, 2006a,b), a full assessment of the role of cryospheric remote sensing in integrated hydrologic observations is beyond the scope of this particular study.

Microwave emissions from the land surface have a strong sensitivity to several hydrologic variables, including soil and snow water content, as well as other land-surface characteristics such as vegetation water content, soil texture, and surface roughness (e.g., Jackson et al. 1999). As such, the development of inversion algorithms for the production of minimally biased estimates of the variable of interest, e.g., soil moisture, from passive and active microwave measurements, has proven challenging. Stated more simply, the variable of interest (such as soil moisture) is not directly observed by the sensor (a problem endemic to many types of environmental sensors) and has to be indirectly estimated using ancillary data. The inversion problem is further complicated by the need to account in algorithm development for the tremendous heterogeneity of land surface within a sensor footprint. Hence, there is a pressing need for focused efforts to develop enhanced algorithms that account for the scaling of relevant heterogeneous variables within satellite footprints to best understand the relationship between observables and derived hydrological products. The integration of different types of data—traditional measurements such as grab samples, in-situ sensor network data, and airborne data—will help in the development, verification, and validation of these algorithms. The problems posed by land-surface heterogeneity underscore the need for terrestrial observatories for hydrologic and related environmental sciences; it is the integration of in-situ aircraft and satellite observations that will ultimately yield significant advances in understanding Earth processes, while also advancing individual sub-disciplines.

Considerable engineering challenges also pose obstacles to advancing space-based hydrological observation. For example, remote sensing of the moisture content of surface soils has been most successful using passive microwave instruments. The capability currently exists to retrieve surface (0-2 cm) soil moisture estimates from the space-based Advanced Microwave Scanning Radiometer-EOS (AMSR-E) instrument at frequencies of about 6 GHz (C band) (Njoku et al., 2003). However, previous research has shown that frequencies in the 1 to 3 GHz (L band) range are better suited for soil moisture retrievals because the microwave emissions emanate from within a deeper soil layer (0-5 cm) (e.g., Jackson et al.,

1995, 1999). Moreover, since biosphere-atmosphere moisture exchanges are influenced by the soil moisture within the entire vegetation root zone, the ability to remotely monitor soil water within the deeper soil profile would be a major advance in Earth observation science. The inherent spatial resolution of AMSR-E soil moisture data is 60 km, yet higher resolution (e.g., 10 km) is optimal for hydrometeorological application. The spatial resolution of microwave data is controlled by antenna size: wider antennas yield higher spatial resolutions. For example, a spaceborne L-band passive microwave instrument would require a 6-meter wide antenna to deliver soil moisture at 40-km spatial resolution. While land-based systems such as ground-penetrating radar offer promise for monitoring deeper soil water content, and airborne L-band passive microwave instruments have proven soil moisture mapping capabilities (Wang et al., 1989; Jackson et al., 1995, 1999, 2002), innovative sensor design and antenna technologies are required before the success of ground and aircraft systems can be achieved in space.

A third avenue for pursuing major advances in hydrologic remote sensing is space geodetic remote sensing. Water cycle observations derived from satellite altimetry (e.g., TOPEX/Jason) and from space-based measurements of Earth's time variable gravity field (e.g., the Gravity Recovery and Climate Experiment, GRACE) provide opportunities to observe terrestrial hydrology in new and informative ways (Alsdorf and Lettenmaier, 2003; Famiglietti, 2004; Lettenmaier and Famiglietti, 2006). Birkett (1995, 1998), Birkett et al. (2002), and Alsdorf et al. (2001) have demonstrated that satellite altimetry can successfully monitor surface-water heights from rivers, lakes, and floodplains. Satellite altimetry missions such as TOPEX/Poseidon and Jason, which are optimized for ocean applications, have great promise for monitoring the elevation and storage changes of inland water bodies, but unless future altimetry missions (Alsdorf et al., 2007) give greater consideration to continental freshwater targets, their utility in hydrology will be limited. The GRACE mission (Tapley et al., 2004) is now providing estimates of total (combined snow, surface water, soil moisture, and groundwater) water storage variations from the large basin to continental scales and providing new insights into the hydroclimatology of terrestrial water storage. When combined with ancillary data, GRACE can also be used to estimate evapotranspiration (Rodell et al., 2004; Ramillien et al., 2005; Swenson and Wahr, 2006) and discharge (Syed et al., 2005) for large (200,000 km^2) river basin systems. While interferometric synthetic aperture radar (InSAR) has been shown to successfully monitor ground deformation resulting from groundwater recharge and discharge (Bawden et al., 2001), the GRACE mission may provide the only viable means for quantifying groundwater storage changes using remote sensing (Rodell et al., 2006; Yeh et al., 2006). An InSAR mission has been proposed by the Decadal Survey (NRC, 2007) for applications such as glacier velocity and surface elevation. With such new opportunities for hydrologic observation comes the challenge to the hydrologic community to embrace these non-traditional data types and to incorporate them into modeling and analysis studies. The need for such methods is highlighted in Chapter 4, for example in "Impacts of Agriculture on Water Resources: Tradeoffs between Water Quantity and Quality in the Southern High Plains," and "Hydrological Observations Networks for Multidisciplinary Analysis: Water and Malaria in Sub-Saharan Africa," where there is a great need for integrative data at large spatial scales or in remote locations.

Beyond the three challenges describe here, several other factors impede progress towards integrating remote observations with in-situ and aircraft data. For example, some critical stocks, fluxes, and properties of the terrestrial water cycle remain poorly monitored. Inadequate spatial and temporal sampling rates, deficiencies in retrieval algorithms, technological obstacles, or simple lack of a dedicated mission, all contribute to gaps in a comprehensive, satellite-based hydrology observing system. For example, while liquid precipitation is well monitored from space, the Tropical Rainfall Monitoring Mission (TRMM) does not provide higher-latitude coverage. Remote sensing of snowfall and snow depth are critical research areas, in particular due to the rugged conditions and inaccessibility of high-altitude basins. Chapter 4's "Mountain Hydrology in the Western United States," elaborates on the pressing need for these measurements. Remote sensing of evapotranspiration remains problematic owing to its reliance on several other variables (e.g., surface radiation and meteorology), many of which are not well measured from space. Satellite monitoring of water quality is an important research frontier.

All of the research areas described above are certain to advance with development of embedded sensor networks, including new sensors, and integration in environmental observatories. It is crucial, however, for researchers working on different systems and scales to keep abreast of advances in related disciplines and to work collaboratively, so that the full potential of spaceborne observations of hydrological sciences can be reached.

In spite of the limitations identified here, great progress has been made towards routine monitoring of certain water fluxes and storages, while continued progress is anticipated from current and upcoming missions. Table 2-3 gives an overview of current and near-future hydrologic remote sensing capabilities. For example, TRMM has been successful in providing precipitation observations that, when combined with other sensor data and model output, form the backbone of the widely-used Global Precipitation Climatology Project (GPCP) products (Huffman et al., 1997). Snow extent, because of its high albedo, has traditionally been well observed by sensors such as the Advanced Very High Resolution Radiometer (AVHRR) and the Moderate Resolution Imaging Spectroradiometer (MODIS) (Hall et al., 2002), while snow water equivalent products, e.g., from AMSR-E (e.g., Kelly et al., 2004), are beginning to emerge. AMSR-E now

TABLE 2-3 Overview of Current and near-future Hydrologic Satellite Remote Sensing Capabilities

Hydrologic Variable	Satellite Sensor	Spatial Resolution[a]	Time Period of Observation
Snow extent	Scanning Multichannel Microwave Radiometer (SMMR)	27 km × 18 km to 148 km × 95 km	1978-1987
	Special Sensor Microwave/Imager (SSM/I)	15 km × 13 km to 69 km × 43 km	1987-present
	Moderate Resolution Imaging Spectroradiometer (MODIS)/Terra	500 m × 500 m	2/2000-present
	Moderate Resolution Imaging Spectroradiometer (MODIS)/Aqua	500 m × 500 m	7/2002-present
Snow water equivalent	Scanning Multichannel Microwave Radiometer (SMMR)	27 km × 18 km to 148 km × 95 km	1978-1987
	Special Sensor Microwave/Imager (SSM/I)	15 km × 13 km to 69 km × 43 km	1987-present
	Advanced Microwave Scanning Radiometer – E (AMSR-E)	6 km × 4 km to 75 km × 43 km	5/2002-present
Surface-water height	European Remote Sensing-1 (ERS-1) Radar Altimeter	Every 320 m along track	1991-2000
	TOPEX/Poseidon	Every 580 m along track	1993-2005
	European Remote Sensing-2 (ERS-2) Radar Altimeter	Every 320 m along track	1996-present
	ENVISAT Radar Altimeter-2	Every 320 m along track	2002-present
	Jason-1	Every 290 m along track	2002-present
	Geosat Follow-on	Every ~ 600 m along track	2000-present
Soil moisture	European Remote Sensing (ERS)-1 SAR	Variable, ~ 20 m × 15,8 m	1991-2000

TABLE 2-3 (continued)

	ERS-2 SARSatellite Sensor	Variable, ~ 20 m × 15,8 m	1996-present
	ENVISAT ASAR	150 m × 150 m	2002-present
	Advanced Microwave Scanning Radiometer-E (AMSR-E)	51 km × 29 km	5/2002-present
Groundwater[b]	Gravity Recovery and Climate Experiment (GRACE)	450 km × 450 km	3/2002-present
Total water storage	Gravity Recovery and Climate Experiment (GRACE)	450 km × 450 km	3/2002-present
Precipitation	Geostationary Operational Environmental Satellite (GOES)	4 km × 4 km or 8 km × 8 km	1978-present
	Special Sensor Microwave/Imager (SSM/I)	25 km × 25 km	1987-present
	Tropical Rainfall Measuring Mission (TRMM/TMI)	0.25° × 0.25°	1998-present
	Advanced Microwave Sounding Unit (AMSU)	4 km × 4 km 16 km × 16 km or 48 km × 48 km	1998-present 1998-present
Evapotranspiration[a]	Moderate Resolution Imaging Spectroradiometer (MODIS)/Terra	500 m × 500 m	2/2000-present
	Moderate Resolution Imaging Spectroradiometer (MODIS)/Aqua	500 m × 500 m	7/2002-present
	Gravity Recovery and Climate Experiment (GRACE)	450 km × 450 km	3/2002-present
Streamflow[a]	From surface-water height data above		
	Gravity Recovery and Climate Experiment (GRACE)	450 km × 450 km	3/2002-present

[a]native resolution; multiple entries imply frequency dependence.
[b]not measured directly; ancillary hydrological data required to derive variable.

provides estimates of surface soil moisture (Njoku et al., 2003), and GRACE produces estimates of monthly variations in total water storage from the large river basin to continental scales (e.g., Syed et al., 2008). Surface-water heights can be monitored from satellite altimeters for several inland water bodies and large rivers, from which discharge estimates can be derived (Alsdorf et al., 2007).

The following three examples demonstrate the role that satellite remote sensing can play as a key element in any strategy to integrate diverse measurement types across space-time scales to provide best-available information to decision-makers and researchers. The focus is on new areas for hydrological remote sensing in order to highlight the types of research, understanding, and products that the satellite information will enable.

Remote sensing and groundwater storage variations. Groundwater accounts for roughly 20 percent of global freshwater consumption. However, accurate monitoring of groundwater storage variations, including recharge and discharge, is a difficult task owing to the sparse nature of groundwater well measurements and the dearth of in-situ soil moisture sensors. Several authors have demonstrated the strong correlation between InSAR observations of land deformation with aquifer compaction (Galloway et al., 1998; Hoffman et al., 2001), with seasonal variations in groundwater storage (Watson et al., 2002), and with groundwater pumping and recharge (Bawden et al., 2001; Lu and Danskin, 2001). More recently, Rodell et al. (2006) and Yeh et al. (2006) have shown that observations of surface and unsaturated water mass can be removed from GRACE estimates of total water variations (in the High Plains aquifer, in the Mississippi basin, and in Illinois) to isolate the groundwater storage change signal. However, these latter studies highlight the need for greatly enhanced soil moisture and surface-water storage observations to produce minimally biased groundwater storage change estimates. Hence, a blueprint for a groundwater observing system in a large aquifer system such as the High Plains would include an embedded sensor network of groundwater monitoring wells, in-situ soil moisture sensors, gravimeters, airborne and remotely sensed soil moisture, and GRACE data, assimilated into a spatially distributed hydrological or groundwater model. The observing system would provide aquifer-average and spatial patterns of groundwater recharge, water table variations, and discharge, which would be available in near-real-time to the user community, using cyberinfrastructure technologies. Without the combination of in-situ sensing, remote sensing constraints provided by GRACE, and the aircraft-satellite-sensed soil moisture, a viable observing system would not be possible.

Remote sensing and the three-dimensional distribution of terrestrial waters. Given the importance of fresh water to the human population, it is surprising to note there is no comprehensive, large-area, freshwater observing system that can describe the lateral and vertical distribution of surface, soil, and groundwater across and through the landscape. The groundwater observing system outlined above can be extended to form a terrestrial waters observing system by the addition of snow and surface-water monitoring. An array of advanced sensors for snow depth (see Chapter 4, Mountain Hydrology in the Western United States for a detailed discussion of the challenges for advancing snow remote sensing) and surface-water storage could be deployed with the soil moisture network for in-situ tracking of surface mass variations. As above, airborne and satellite soil moisture remote sensing would provide critical boundary conditions. A hydrology-specific satellite altimetry mission is an essential component in this framework, for without it there is no way to routinely monitor seasonal variations in surface-water storage. Again, a data assimilating model could serve as the integrator for the various data types, and its output would include maps of snow, surface, soil, and groundwater storage.

The National Operational Hydrologic Remote Sensing Center's (NOHRSC) effort to integrate airborne and ground-based snow survey data with a weather-forecast model to estimate snow water equivalent provides an example of how this can be accomplished (http://www.nohrsc.nws.gov). A three-dimensional characterization of terrestrial water distribution would be an invaluable contribution to water management and hydrologic research. Moreover, it would form a first step towards characterizing the three-dimensional circulation of surface and subsurface waters, which is critical for understanding water and contaminant flowpaths.

Remote sensing and water quality along coastal margins. Over half of the U.S. population resides in coastal counties, where most rapid growth rates in the country are occurring (Crossett et al., 2004). Unfortunately, degradation of land, air, and water resources is a consequence of the continued urbanization occurring in these regions. Remote sensing of terrestrial and coastal water quality along land-ocean margins is an important frontier in satellite monitoring of environmental quality. In fact, it is now well documented that the quality of terrestrial freshwater inputs into the coastal zone is a critical determinant of beach water quality (Ahn et al., 2005).

If freshwater remote sensing methods can be enhanced, they could be combined with satellite observations of precipitation, soil moisture, river heights, coastal ocean water quality, and existing sensor web technologies for in-situ monitoring in both land and coastal waters. The satellite data could serve as pods in the sensor web, triggering intensive sampling in regions where precipitation is actively falling or anticipated, or as regional soil moisture or river heights increase and contaminants are mobilized, or with the spreading of a freshwater or coastal plume. In Southern California for example, where the beach water quality sampling takes a full day, results are often obsolete by their time of release (Jeong et al., 2006). Consequently, millions of beachgoers are either put at risk from late warnings, or are inconvenienced by unnecessary closures. A land-ocean margin water-quality observing system would integrate sensor web data from river, groundwater, and wastewater systems along with airborne and satellite fresh and coastal ocean water quality. A coupled, regional land-ocean model would assimilate these data and provide real-time forecasts of beach water quality, greatly enhancing capabilities for timely beach closure. As in the examples above, the synoptic view provided by remote sensing would be essential for integrating sensor web data into a viable scheme for real-time prediction along large areas of densely populated coastline.

The Southern California Coastal Observing System (SCCOS); (http://www.sccoos.org) provides a good example of how such a system would function, but primarily for the ocean. The Real-Time Coastal Observation Network (ReCON; Ruberg et al., 2007) offers a second example for the North American Great Lakes, but as with the SCCOS, for the water body only, without significant monitoring and modeling of the contributing watersheds. Conversely, the South Florida Water Management Case Study describes a comprehensive, advanced terrestrial surface-and groundwater monitoring system that could be readily linked to an ocean observing system like SCCOS or ReCON. Remote sensing of surface-water heights, inundation extent, and groundwater storage changes could be integrated into existing surface- and groundwater monitoring and modeling activities in South Florida, and coupled to a coastal observing framework. Such an integrated observing system would not only greatly improve inland and coastal water-quality monitoring and prediction, but would enhance decision-making information streams with important implications for balancing economic development and societal demands for wetland restoration in the Everglades region.

SENSOR MAINTENANCE

It should be clear from the discussions in this chapter that a major issue is the cost and maintenance of sensors at all scales. This is not only true for sophisticated chemical sensors; even off-the-shelf technology that senses fairly basic parameters (e.g., turbidity, chlorophyll, dissolved oxygen, oxidation-reduction potential, and temperature) can pose this challenge. Maintenance problems can easily consume any perceived cost savings of technologically advanced sensors, and few entities may be interested in providing financial support for ongoing operations and maintenance of a network. Operation and maintenance costs may even be a critical and understated problem in the National Science Foundation (NSF) Environmental Observatory programs, and NASA and NOAA satellite programs have not been immune to the problem either. Therefore, observations programs at any scale may wish to carefully consider these costs at an early stage of project planning.

COMMUNITY INVOLVEMENT

For certain types of environmental pollutants such as lead and arsenic, as well as pathogens and diseases such as malaria, communities can become important components of sensor networks, empowering individuals to assist scientists with monitoring and control. For example, in the Chapter 4 case study on "Water and Malaria in Sub-Saharan Africa," community members are being trained to be active participants in a study of monsoons and malaria in the semiarid Sahel zone of Africa. This group is studying the environmental determinants of malaria outbreak using observations, and numerical simulation, including hydrologic modeling. Direct community involvement has also proved effective in measuring lead contamination in the inner-city and arsenic contamination in India and Asia, using test kits.

In the United States, especially in the eastern states, citizens have formed numerous watershed groups. Their purpose is to help to maintain the environmental health of the streams or to work to recover the health of impacted streams. One activity that is common in these groups is the environmental monitoring of the rivers. Often they are supported and advised by nongovernment and government organizations. For example, the Northern Virginia Soil and Water Conservation District provides training and equipment to volunteers in the assessment of ecological conditions in streams based on the presence and abundance of benthic macroinvertebrates. Volunteers also take chemical measurements such as pH, total dissolved solids, temperature, discharge, nitrate/nitrite, and turbidity.

Experienced monitors, often including members of the scientific community, train volunteers to implement appropriate methodologies for aquatic monitoring. Attention to such detail is important because some methodologies (e.g., for sampling benthic macroinvertebrates) are region specific. Training volunteers makes it possible to study the streams more frequently and to have more monitoring stations. These types of activities can be a mechanism leading to more comprehensive studies and conscious involvement of the community in environmental issues.

A long-standing citizen-based observation program is the Cooperative Observer Program of the National Weather Service (NWS). This is a network of over 11,000 volunteers, established in 1890, to make weather observations and establish and record climate conditions in the United States, with a traditional emphasis on agriculture. Today the network is increasingly used to support meteorological and hydrological forecasts and warnings and to verify forecasts.

Making good use of active community involvement in sensing requires careful training and validation of results, as well as detailed calibration of test kits against traditional laboratory-based analysis. Nevertheless, such community involvement expands scientific databases when integrated with other traditional and developing methods in the hydrologic and related sciences.

SUMMARY

This chapter summarizes current and emerging sensor and sensor networking technologies that are being developed for measuring hydrological and environmental processes. The research and development activities related to embedded sensor networks, biogeochemical sensor technologies, and at larger scales sensors designed for airborne and space platforms, offer significant opportunities to advance our understanding of critical hydrological and environmental processes through improved observations at different scales.

The committee heard a number of presentations from federal agency scientists, academics, and private industry. These presentations ranged from descriptions of current measurement approaches (e.g., NWS snow measurement techniques in the Sierra Nevada that have not changed appreciably for 100 years) to applications of sensors networks (e.g., the real-time control storm runoff).

From these presentations and related material reviewed by the committee, emerged Figure 2-10, which describes the participants and summarizes the steps

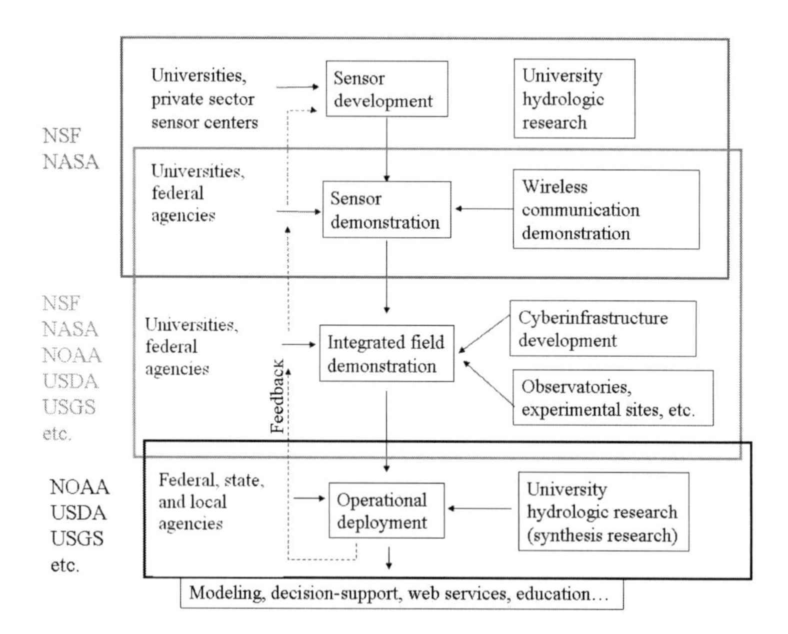

FIGURE 2-10 A summary of the steps involved in advancing from experimental development of a sensor through its operational deployment. The role of different, major activities and participants are shown as contributing to the process at different stages, and the federal agencies that would have major interest in these stages are also shown. USDA = U.S. Department of Agriculture.

in going from sensor development to operational deployment. There are many players involved in each step, and Figure 2-10 attempts to list them, and much on-going activity. Nonetheless, the process is far from smooth or seamless. As examples, current operational sensor networks generally have no plans to utilize new sensor and sensor networking technologies that would integrate various types of measurements to produce improved operational observations. At the same time, most university centers working on sensor development have no plans or resources to test their sensors within larger integrated field demonstrations that incorporate a variety of sensors, inter-connected through embedded networks and cyberinfrastructure for their potential delivery to data users.

From Figure 2-10, the committee sees Balkanization of the sensor development process, and subsequent gaps in agency programs, including potentially new roles for NSF as the nation's primary research funding agency. Specific recommendations are offered in Chapter 5 for addressing these gaps.

3

Integrating Observations, Models, and Users

As presented in Chapter 2, rapidly increasing sensing capacity is greatly expanding our ability to measure and monitor the conditions and processes that are critical to understanding and managing hydrologic systems. These technical advances are coming at a time when the demand for new and more accurate hydrologic information is rapidly increasing.

Models are typically used to extend the utility of individual measurements. They can be used to interpolate point data, integrate point and remotely sensed data, and estimate unmeasured quantities. Models can also be used to forecast future conditions, hindcast past conditions, and simulate hypothetical conditions, such as hydrologic states and fluxes under alternative future conditions. Further, models are also useful for designing measurement and monitoring systems, as they provide information on the value of specific measurements and can be used to determine the types, amount, and geospatial distribution of data that need to be collected. Real-time modeling of ydrologic systems offers an exciting opportunity for integrating data and models through "data assimilation", an approach that is being used successfully in weather forecasting.

There is wide variety of models that are used in the hydrologic community, for example, models for predicting water quality, flood forecasting, water supply, and so forth. They have a wide range of temporal and spatial scales (e.g., spatially distributed sub-diurnal hydrologic models, basin-averaged hydrologic models, and regional total maximum daily load (TMDL) based non-point pollution models). Thus, how observations are merged with model predictions, and the usefulness of using models to evaluate observational networks, will be model- and application-oriented, and is a focus of research. Overall, model development has far exceeded our ability to provide these models with field data. Rather than supply a comprehensive review of the range of model applications that would find the integrated observations useful, a cross-section of models is presented through the specific case studies presented in Chapter 4. The committee expects that integrated programs like the National Ecology Observatory Network (NEON) and the Water and Environmental Research Systems (WATERS) Network will offer additional research results showing the benefits of integrating observations and models.

Neither data nor models have value unless they are used. And they can only be used if they can be easily discovered, acquired, and understood in a timely and convenient manner by those who wish to apply them to practical issues such as flood forecasting, water availability modeling, and ecological flows, as inputs to decisionmaking. The communication and delivery of data and information (including their interpretation, quality, and uncertainties) to such end-users is the back-bone to a beneficial integrated system.

Therefore, efficient use by society of these new sources of data from land, air, and space requires concomitant improvements in the capture and archiving of these data, in the modeling of hydrologic systems, and in the communication of hydrologic data and information to researchers, water managers, and other users. However, achieving these goals requires a level of cyberinfrastructure not currently available or even designed. There are critical and extensive cyberinfrastructure needs if models are to routinely and efficiently take advantage of advances in measurements. It is critical that cyberinfrastructure evolve in concert with new developments in sensing capacity and hydrologic modeling.

Thus, this chapter first presents new opportunities in the merging of observations with models, followed by a discussion of the cyberinfrastructure needed to support these models and their application to societal needs.

APPROACHES FOR INTEGRATING OBSERVATIONS AND MODELS

Real-Time Environmental Observation and Forecasting Systems

The availability of tremendous computational power coupled with widespread communication connectivity has fueled the development of real-time environmental observation and forecasting systems. These systems offer the opportunity to couple real-time in-situ monitoring of physical processes with distribution networks that carry data to central processing sites. The processing sites run models of the physical processes, possibly in real-time, to predict trends or outcomes using on-line data for model tuning and verification. The forecasts can then be passed back into the physical monitoring network to adapt the monitoring with respect to expected conditions (Steere et al., 2000). Both wireless networks of sensors and sensors webs will enhance this development.

This approach is being successfully used in weather forecasting. Using protocols developed around the Global Telecommunication System (GTS), the National Weather Service has created an integrated network that interconnects meteorological telecommunication centers with point-to-point and multi-point circuits. This allows national weather centers to obtain global data from in-situ networks and environmental satellites that are "assimilated" into their forecast models.

A similar vision can be advanced for prediction in hydrologic systems, from daily water-quality forecasts in major estuaries and near-shore regions to timing of snowmelt runoff for hydropower scheduling to flood forecasting to operation of irrigation drainage facilities to protect stream water quality. While selected federal and state agencies are starting to move in this direction (e.g., the National Oceanic and Atmospheric Administration's [NOAA] Drought Information Center, and its products), most are not. Opportunities for advances across the water sector are enormous. While the models can be used to help "design" data collections systems—the so-called operational sensor simulation experiments (OSSEs)—the more fundamental challenge is developing and testing methods for merging the diverse data from a variety of sensors into existing models.

Data Assimilation

As an example of the potential for integrating real-time measurements into predictive tools and thence into decision-support systems, we review the historical context and some recent challenges in hydrologic data assimilation. The merging of multiscale observations and models when both observations and model predictions are uncertain is referred to as data assimilation. Data assimilation has been widely used in atmospheric and ocean sciences (e.g., Bennett, 1993; Evensen, 1994), and is central to operational weather forecasting, in which a variety of observed weather states from disparate sensors are used to update the model states, and subsequently forecasts. In hydrology, research on data assimilation procedures has a relatively long but sparse history, going back to the 1970s. That earlier work generally focused on simple linear or linearized models within a Kalman filtering framework (see Wood and Szollosi-Nagy, 1980) and was applied to problems such as flood forecasting.

Major limitations of the early work included difficulties in handling the highly nonlinear nature of surface hydrologic systems, especially during intense storms, the general absence of observations of key states, like soil moisture, and computational limitations. The application of more advanced techniques is a relatively new phenomenon in hydrology. The renewed interest in hydrologic data assimilation has been spurred in part by the increased availability of remote sensing and ground-based observations of hydrologic variables and/or variables (like soil moisture and surface temperature) that can be related to surface hydrologic processes, and along with improved computational power (Houser et al., 1998; McLaughlin, 1995, 2002; Reichle et al., 2001a;b; Crow and Wood, 2003).

The basic objective of data assimilation is to better characterize the state of the hydrologic or environmental system, where information sources include process models, remote sensing data, and in-situ measurements. Until recently, research on data assimilation in land-surface hydrology was limited to a few one-dimensional, largely theoretical studies (e.g., Entekhabi et al., 1994; Milly, 1986), primarily due to the lack of sufficient spatially distributed hydrologic observations (McLaughlin, 1995). However, the feasibility of synthesizing distributed fields of surface states (e.g., soil moisture) by the novel application of four-dimensional data assimilation (4DDA) within the construct of a dynamic hydrologic model was only quite recently demonstrated (Houser et al., 1998). More recently, researchers in hydrology (e.g., Margulis et al., 2002; Crow and Wood, 2003; Wilker et al., 2006) have exploited both aircraft remote sensing observations from intensive field campaigns like SGP97 and SGP99 over the Southern Great Plains (SGP) domain and spaceborne observations.

Encouraging though these short-term demonstrations may be, the area is still quite limited relative to the continental and global domains for which remote sensing data sets from the suite of the National Aeronautics and Space Administration (NASA) earth observation system (EOS) satellites are available. At these scales, computational and algorithmic issues need to be better developed to integrate multiscale observations and models (Zhou et al., 2006; McLaughlin et al., 2006). Therefore, it is recognized that applying data assimilation to provide better integration of observations—from sensor pods from embedded networks to operational sensor webs and from local high-resolution process models to continental-scale, macroscale "earth-system" models—will represent a significant scientific challenge that necessitates investment in research and demonstration projects by the National Science Foundation (NSF), NASA and operational agencies to fully utilize the potential from the observational and modeling capabilities that currently exist.

A way forward for the research in merging observations and modeling, both which exist at a variety of scales, is to apply a scale-recursive assimilation/smoothing procedure (Basseville, 1992; Daniel and Willsky, 1997; Luettgen and Willsky, 1995; Gorenburg et al., 2001.) Figure 3-1 shows the multiscale problem schematically, where a given measurement (or model output, or sensor) may provide information at other scales. An in-situ sensor within an embedded network at scales smaller than the finest grid may provide spatial information for the larger scale as well as information regarding sub-grid variability.

Multiscale approaches can be used with any of the covariance-based data-assimilation algorithms, such as optimal interpolation, three-dimensional variational (3-D Var) approaches, and Kalman filtering and its derivatives such as Extended Kalman filtering or Ensemble Kalman filtering. Additionally, multiscale methods offer a feasible computational approach to one of the largest potential challenges of widespread embedded sensor networks: the production of large

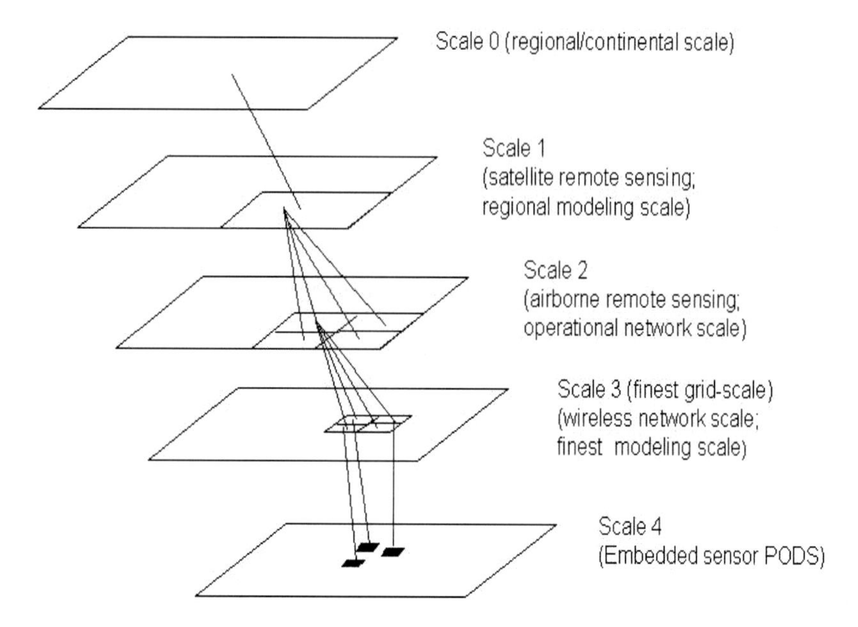

FIGURE 3-1 Portion of an inverted tree showing potential scales and measurements. Alternatively the scale 1 node has a parent (scale 0) and four children at scale 2.

amounts of data that could be overwhelming if not properly managed. Hence, the embedded computational ability of embedded sensor networks, combined with multiscale merging with coarser-scale observations and data can allow predictions across scales.

CYBERINFRASTRUCTURE: MANAGING THE DATA AND DELIVERING THE PRODUCTS

Cyberinfrastructure was defined broadly by an NSF Blue Ribbon Panel (2003) as:

> The base technologies underlying cyberinfrastructure are the integrated electro-optical components of computation, storage, and communication that continue to advance in raw capacity at exponential rates. Above the cyberinfrastructure layer are soft

> ware programs, services, instruments, data, information, knowledge, and social practices applicable to specific projects, disciplines, and communities of practice. Between these two layers is the cyberinfrastructure layer of enabling hardware, algorithms, software, communications, institutions, and personnel.

Cyberinfrastructure encompasses the coordination and deployment of information technologies that integrate observations and measurements, high-performance computing, management services, visualization services, and other advanced communication and collaboration services in a networked environment. Moreover, cyberinfrastructure necessarily includes the human resources necessary to support research and applications. In the remainder of this section, we discuss some of the current cyberinfrastructure challenges, present case studies illustrating promising developments in cyberinfrastructure, and conclude with a comprehensive vision for cyberinfrastructure that enables real-time environmental observation and forecasting.

Current Challenges

There are many cyberinfrastructure challenges, especially as they relate to communication technologies, associated with developing wireless networks and the integrated, seamless, and transparent information management systems that can deliver seismic, oceanographic, hydrological, ecological, and physical data from sensors to a variety of end users in real time. One example of a research project that has advanced the communication and real-time management of distributed, heterogeneous data streams from sensor networks in the San Diego region through the confluence of several cyberinfrastructure technologies is ROADNet (Real-time Observatories, Applications and Data management Networks; https://roadnet. ucsd.edu; Woodhouse and Hansen, 2003; Vernon et al., 2003). In particular, ROADNet employs (1) commercial software for data flow, buffering and distribution, data acquisition, and real-time data processing; (2) open-source solutions for data storage (i.e., Storage Resource Broker), integration, and analysis (i.e., Kepler Workflow System); and (3) web-services tools for rapid and transparent dissemination of results. ROADNet is used as the middleware and real-time data distribution system for the wireless, real-time sensor network at Santa Margarita Ecological Reserve in southern California (http://fs. sdsu.edu/kf/reserves/smer/).

Developing environmental observatories will increasingly need to focus on developing the technologies that can better enable users to use and understand data from sensor webs. It will be especially important to provide Internet access to integrated data collections along with visualization, data mining, analysis, and modeling capabilities. Heterogeneities in platforms, physical location and naming of resources, data formats and data models, supported programming interfaces, and query languages should be *transparent* to the user, and cyberinfrastructure will need to be able to adapt to new and changing user requirements for data and data products. Cyberinfrastructure tools to achieve this are the major challenge in the integration of hydrologic observations, and the use of these observations by both the research and applications communities.

Large-scale environmental observatories like EarthScope, the WATERS, NEON and others will consist of hundreds to thousands of distributed sensors and instruments that must be managed in a scalable fashion. Cyberinfrastructure must enable scientists to remotely query and manage sensors and instruments, automatically analyze and visualize data streams, rapidly assimilate data and information, and integrate these results with other ground-based and space-based observations. With the advent of highly distributed embedded networked sensors, along with potential and real advances in biogeochemical sensors, the cyberinfrastructure will become even more important. Development of automated quality assurance and quality control procedures, adoption of comprehensive metadata standards, and the creation of data centers for curation and preservation are critical for the success of environmental observatories and the longevity and usability of their data holdings. Data are not useful unless they are high quality, well organized, well documented, and securely preserved yet readily accessible.

Achieving this will require a major paradigm shift from the way most hydrologic data are now handled by both the research and applications communities. Data must go directly from sensors to a data and information system, with quality assurance and quality control done within the system. Automatic quality control algorithms need to be built in, but the system must also allow efficient data processing by scientists responsible for the data. The paradigm shift is that data will not go to an investigator's computer to be processed and later submitted to a data system. Rather, data will go directly into a data system, with the responsible investigator involved in immediate, timely data processing facilitated by the data system.

Emerging environmental observatories will clearly depend upon cyberinfrastructure to reduce the significant costs in money and people required to manage distributed sensing resources. Importantly, incipient environmental observing systems may represent a significant market presence that can encourage creation and acceptance of industry standards for sensor compatibility, communication, and sensor metadata. For instance, standardized approaches that automate capture and encoding of sensor metadata can facilitate the process whereby sensor data are ingested, quality assured, transformed, analyzed, and converted into publishable information products (Michener, 2006). Automatic metadata encoding should also enable scientists to track data provenance throughout data processing, analysis, and subsequent integration with other data products.

Environmental observing systems require data center and web services approaches that enable different users to specify data and information that can be automatically streamed to their computer. Furthermore, it should be possible to specify alternative views to data from raw through processed (quality assured) through integrated and synthesized information (e.g., graphic summaries of model outputs) depending on scientific need and level of user experience.

A final challenge may be the most difficult of all. This report notes in various places that integrating measurements in multiple sciences (e.g., biology, hydrology, and meteorology) is challenging, but overcoming this challenge is non-trivial. For example, simply gathering the data in a web-based portal (see the applications examples below) does not solve the problem. Research is needed in how cross-disciplinary information is communicated and exchanged. For example, the discipline of spatial ontology attempts to find a rigorous set of terminology for the same phenomena and geographic features in different disciplines. A serious attempt will have to be made to build cross-disciplinary databases if truly integrated information systems are to be achieved. The suggestion made in NRC (2006c) was to place the various NSF environmental observatory programs under a parent entity, which would be responsible for cyberinfrastructure development, among other shared activities. This would be a key forum for such discussions.

Examples of Applications

A detailed development of cyberinfrastructure needs is beyond the scope of this report. Here, to illustrate some of the above-mentioned issues in context, we highlight three recent and relevant advances in environmental infrastructure: (1) a "cyberdashboard" for the current EarthScope USArray real-time infrastructure, (2) a community observations data model and corresponding suite of web services for hydrologic applications, and (3) a specific application to flash flood emergency management.

A "Cyberdashboard" for the EarthScope USArray Project

The EarthScope USArray project seeks to study the seismic tomography of the continental United States. As with other networks, significant human effort is required to configure, deploy, and monitor the thousands of sensors that are being constantly deployed and redeployed and to manage their real-time data streams. This process required administrators to log in to multiple computers, edit configuration files, and run executables to properly integrate the new equipment into the existing sensor network.

Cotofana and others (2006) recently applied a services-oriented architecture (SOA) approach to create a "cyberdashboard" for the current USArray real-time infrastructure (Figure 3-2). The architecture included (1) a layer of command-and-control web services, capturing and exposing resource management behaviors of the existing sensor network middleware used by USArray; (2) web-based management applications that orchestrate these services to automate common management tasks; and (3) geographical information system (GIS) capabilities provided by Google Earth™ to display the sites being configured and their surrounding environments.

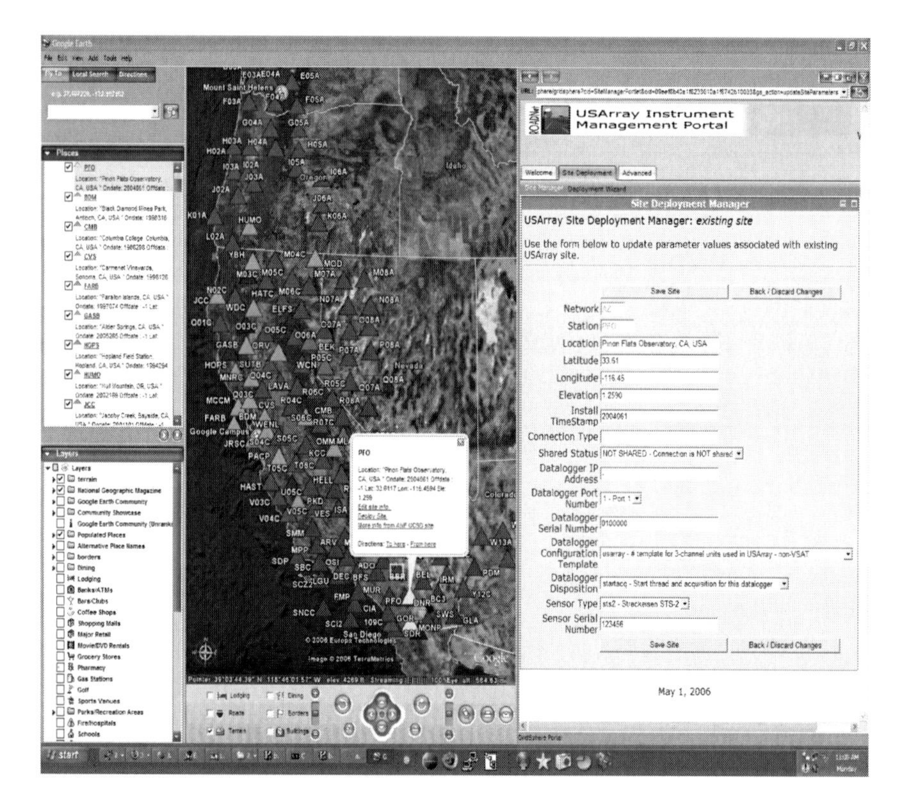

FIGURE 3-2 USArray instrument management cyberdashboard, which provides an intuitive and comprehensive view into system status and operations, as well as control functions over various system resources such as data streams, instruments, data collections, and analysis and visualization tools. SOURCE: Reprinted, with permission, from T. Fountain, San Diego Supercomputer Center.

The cyberdashboard centralizes management activities into one consistent interface, decoupled from the actual systems hosting the underlying sensor network middleware. The management tools also provide a means of keeping track of multiple instrument sites, as they go through the various deployment steps, and automatically check the constraints, reducing input errors. Furthermore, these tools guide the administrators through the configuration of new sites, ensuring that all steps are properly completed in the right order. GIS tools for sensor networks facilitate a number of administration tasks, ranging from visual verification of site coordinates to the planning of new deployments given natural environmental conditions. The cyberdashboard combines all of these elements into an integrated user interface of monitors and controls for observing system management.

A Community Observations Data Model and Web Services for Hydrologic Applications

Efforts undertaken as part of the Consortium of Universities for the Advancement of Hydrologic Science, Inc.'s (CUAHSI) Hydrologic Information System are streamlining access to information from diverse data archives. The Hydrologic Information System (HIS; http://www.cuahsi.org/his) is one component of CUAHSI's mission and comprises a geographically distributed network of hydrologic data sources and functions that are integrated using web services so that they function as a connected whole. The goal of HIS is to improve access to the Nation's water information by integrating data sources, tools, and models that enable the synthesis, visualization, and evaluation of the behavior of hydrologic systems. This will be achieved through a distributed service-oriented system. Significant parts of HIS have been prototyped, and others are under development. Two contributions from HIS are highlighted here: (1) the community observations data model and (2) a suite of web services called WaterOneFlow.

The CUAHSI community observations data model is a standard relational database scheme for the storage and sharing of point observations from a variety of sources both within a single study area or hydrologic observatory and across hydrologic observatories and regions. The observations data model is designed to store hydrologic observations and sufficient ancillary information about the data values to provide traceable heritage from raw measurements to usable information, allowing data values to be unambiguously interpreted and used. The observations data model is an atomic data model that represents each individual observation at a point as a single record in the values table. This structure provides maximum flexibility by exploiting the ability of relational database systems to query, select, and retrieve individual observations in support of diverse analyses.

Although designed specifically with hydrologic observations in mind, this data model has a simple and general structure that can accommodate a wide range of other data from environmental observatories and observing networks, or even model output values. The fundamental basis for the data model is the data cube (Figure 3-3) where a particular observed data value (D) is located as a function of where in space it was observed (L), its time of observation (T), and what kind of variable it is (V). Other distinguishing attributes that describe observations serve to precisely quantify D, L, T, and V. The general structure of the data model, with comprehensive observation metadata, provides the foundation upon which HIS web services are built.

A critical step not yet designed is how to populate with data at different levels of processing, from raw sensor data to higher-level geophysical products, and how to efficiently build quality assurance and quality control into the system. A review of on-line data systems in related fields will show many times over that when research data pass to an individual investigator's desktop for processing en route to an archive, only selected data actually end up in that archive. Cyberinfrastructure needs to capture data automatically, and make it painless for investigators to do quality assurance and quality control.

The second contribution from HIS is a suite of web services, called WaterOneFlow, that enable HIS users to access data, tools, and models residing on different servers. These web services can be invoked from a web browser interface or from various applications or programming environments commonly used by hydrologic scientists. An important function of web services is to facilitate analysis of national hydrologic data in the applications and analysis environment of a user's choice, thereby minimizing the requirement for additional learning by users and providing greatest flexibility for the use of hydrologic data. This is achieved through the reliance on World Wide Web Consortium web services standards. WaterOneFlow presently accesses servers maintained by national water data centers (e.g., U.S. Geological Survey (USGS), Ameriflux, DayMet), and is supported by HIS servers at the San Diego Supercomputer Center (SDSC).

Figure 3-4 depicts the architecture of the WaterOneFlow system. This shows how web services serve to integrate and standardize the access to data from multiple sources such as USGS, National Climate Data Center, observatory servers, and HIS servers that provide data in the format of the observations data model. The primary function of the web portal interface depicted at the top is data discovery and preliminary data exploration. However, once a user has found what data are available and wants to do analysis it is generally more efficient to discontinue using the browser and access the data directly from the working environment of the user's choice. This is depicted in the box on the right in Figure 3-4.

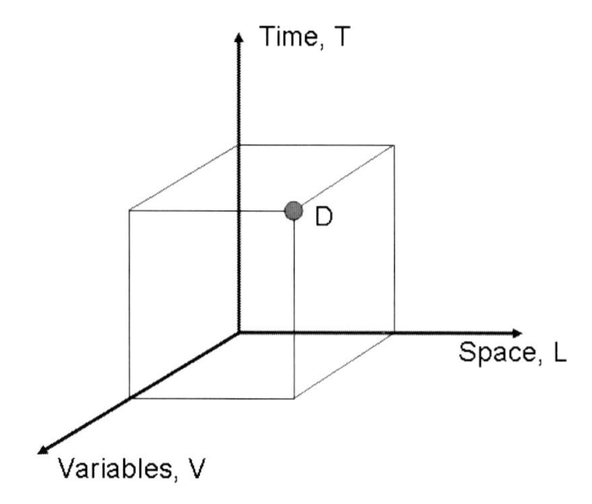

FIGURE 3-3 A measured value (D) is indexed by its spatial location (L), its time of measurement (T), and what kind of variable it is (V). SOURCE: Maidment (2002). © 2002 by ESRI Publishers.

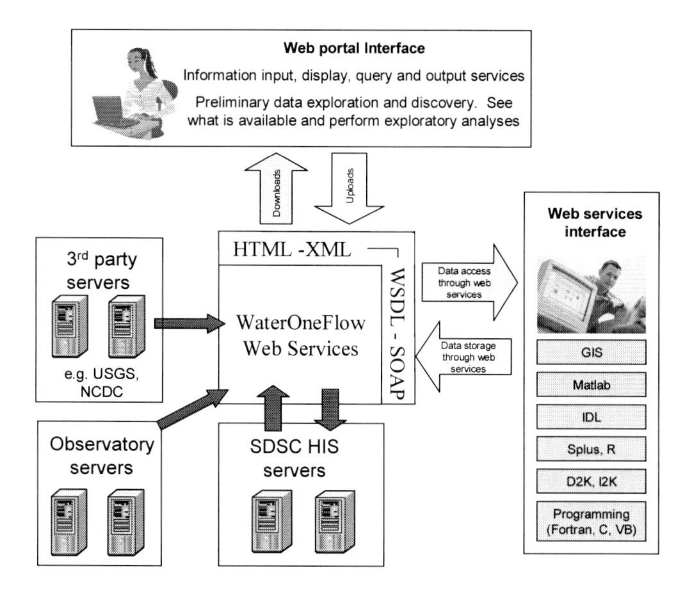

FIGURE 3-4 WaterOneFlow web services architecture. SOURCE: Reprinted, with permission, D. Tarboton, Utah State University. Available on-line http://www.cuahsi.org/his/webservices.html.

Flash Flood Emergency Management through Web Servers

Flash floods have the dubious distinction of resulting in the highest average mortality (deaths/people affected) per event among natural disasters (e.g., Jonkman, 2005). Early warning systems are the means for producing localized and timely warnings in flash flood prone areas, which in many cases are remote, necessitating the use of remotely sensed data. An approach to flash flood warning that has gained acceptance and is used in operations is to produce estimates of the amount of precipitation of a given duration that is just enough to cause minor flooding in small streams over a large area. These *flash flood guidance* estimates are then compared to corresponding now casts or short-term forecasts of spatially distributed precipitation derived from remote and on-site sensor data and numerical weather prediction models to delineate areas with an imminent threat of flash flooding and to issue warnings and mobilize emergency management services (e.g., Sweeney, 1992). Flash flood guidance estimates are derived using GIS information and distributed hydrologic modeling.

Flash flood guidance is not a forecast quantity; rather, it is a diagnostic quantity (e.g., Carpenter et al., 1999; Georgakakos, 2006; Ntelekos et al., 2006). As such, its use for the development of watches and warnings requires assessment of a present or imminent flash flood threat or a possible flash flood threat in the near future (up to six-hours of lead time). As flash floods are local phenomena developing rapidly, it is best to have these assessments made by local agencies that are familiar with the response of the local streams and have access to last-minute local information (be it a phone call from local residents or information from a local automated sensor). Thus, even though a regional center may be producing flash flood guidance estimates over the region with high resolution, effectively assimilating all available real-time data, appropriate means for communication and interaction with such estimates are necessary to allow local agency assessments pertaining to issuing flash flood watches and warnings (a flood is occurring or will occur imminently) and/or taking steps for emergency management. The World Wide Web offers a very effective primary means for communicating and interacting with spatially distributed flash flood guidance estimates and associated precipitation data through client-server arrangements between the regional center and the local forecast and emergency management agencies. The integration of observations, high-performance computing, data management and visualization services, and user collaboration services promises to contribute significantly toward the reduction of life loss from flash floods.

One of the first examples of the use of the web to enhance the effectiveness of producing warnings and emergency management for flash floods is the Central America Flash Flood Guidance System (CAFFG) that has served all seven countries in Central America since the summer of 2004. The system was developed by the Hydrologic Research Center in collaboration with NOAA and country meteorological and hydrologic services with funding from the U.S. Agency for International Development. A regional center in San José, Costa Rica, receives and ingests in real-time all the remotely sensed (satellite rainfall) and on-site operational data for the region. The system currently relies on NOAA estimates of satellite rainfall based on geostationary satellites with real-time assimilation of surface rainfall observations from a variety of conventional, operational-automated sensors in various countries of Central America. Current data-delivery latencies of rainfall estimates from satellite microwave data prevent the use of such information for the development of timely warnings. Other real-time data from World Meteorological Organization-network reporting surface stations are also ingested and used by the models. Uncertainty in observations and models is transformed into uncertainty measures characterizing the nowcasts of flash flood occurrence.

The regional center in Costa Rica hosts the system computational servers used for the production of flash flood guidance with spatial resolution of about 200 km^2 on the basis of hydrologic and geomorphologic principles. A secure web site at the regional center produces a series of guidance products in real time (hourly) for in-country hydrologic and meteorological services (Figure 3-5). These in-country services can download GIS products locally for further processing, possibly update the downloaded information with local data, and generate flash flood warnings and watches. Emergency management agencies and other organizations (United Nations World Food Programme) have access to the web products for planning their deployment activities in areas where high flash flood threat is estimated. Real-time products with coarser spatial resolution that have not undergone in-country meteorological and hydrologic services quality control are disseminated through the World Wide Web for public information (http://www.hrc-lab.org/right_nav_widgets/realtime_caffg/index.php). A multi-month training program was established for regional center staff and in-country users for effective use of the operational system. In addition to system operations and guidance, product interpretation, and use, training also covered methodologies for the recording of flash flood occurrence in small streams to enhance regional observational databases and for the validation of the warnings issued on the basis of CAFFG.

While initial validation results for the system are promising, it is but a first step toward the solution of the flash flood warning problem in ungauged areas. Fruitful areas for improvement are the reduction in the latency of microwave satellite rainfall data for use in the production of flash flood warnings; the deployment of low-cost and maintenance sensors for precipitation, temperature, and flow stage and discharge in remote areas, and the enhancement of the existing cyberinfrastructure to accommodate these; and the improvement and in some cases the development of regional communication networks relying on the World Wide Web for more effective cooperation and data management of the

CAFFG Dissemination Server
Web Access GUI – National Data Interface

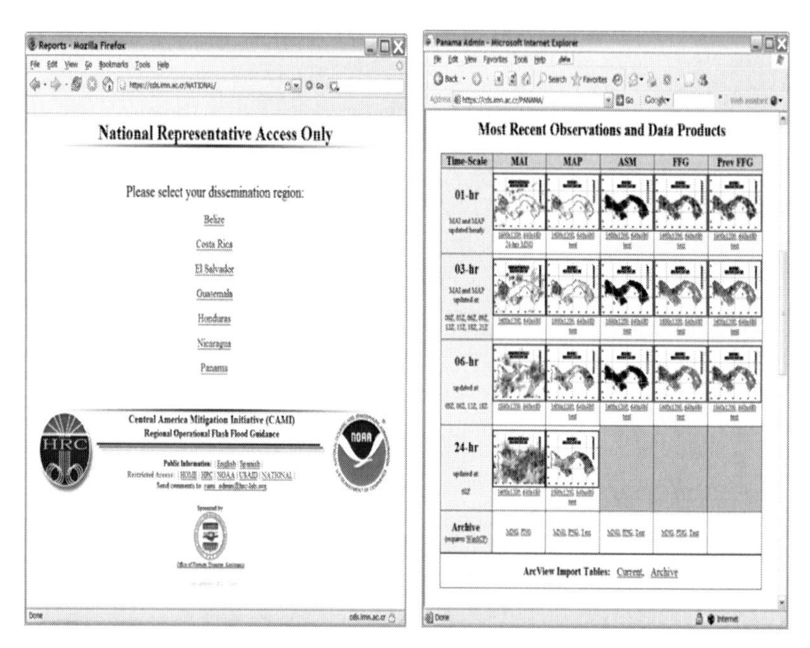

FIGURE 3-5 The secure web site interface of CAFFG. Information and products, such as mean areal precipitation and flash flood guidance, are provided on a country basis for the seven countries of Central America. CAFFG system information may be found in Sperfslage et al. (2004). Reprinted, with permission, from Sperfslage et al. (2004). © 2004 by Hydrologic Research Center.

national meteorological and hydrological forecast and management agencies of the region.

A Vision for Cyberinfrastructure

A future vision for cyberinfrastructure that addresses many of the challenges associated with developing real-time environmental observation and forecasting systems includes automated capture of well-documented data of known and consistent quality from both embedded network systems and air- and space-based platforms; active curation and secure storage of raw and derived data products; provision of algorithms, models, and forecasts that facilitate the integration of data across scales of space and time, as well as the generation of predictions and forecasts; and the timely dissemination of data and information in forms that can be readily discovered and used by scientists, educators, and the public. Realizing this vision will require significant advances in research and development. In particular, new approaches for data quality assurance and quality control, automated encoding of metadata with data, and algorithms that enable integration of data across broad scales of space (single-point sensors to regional-scale hyperspectral imagery) and time (fractions of seconds, in the case of some sensors, to days to weeks, in the case of some remotely sensed data products).

We propose several guiding principles that would facilitate the creation of the cyberinfrastructure needed to support environmental observation and forecasting. First, *open architecture* solutions are central to enabling the rapid adoption of new hardware and software technologies. Second, nonproprietary and, ideally, *open-source* software solutions (e.g., middleware, metadata management protocols) promote the modularity, extensibility, scalability, and security that are needed for observation and forecasting. Third, *development and adoption of community standards* (e.g., data transport, quality assurance/quality control, metadata specifications, interface operations) will more easily support system interoperability. Fourth, *open access to data and information* and the provision of customizable portals are key to meeting the needs of scientists, educators, and the public for timely access to data, information, and forecasts.

In this chapter we have discussed new opportunities for incorporating modeling and data communication into integrated hydrologic measurement systems, and summarized the elements that might go into such a system. In the next chapter we present summaries of case studies developed by the committee based on existing and proposed measurement systems. The purpose of these examples is to provide context for the ideas discussed above and to provide insight into the requirements and challenges associated with the development of integrated hydrologic measurement systems.

4

Case Studies on Integrated Observatories for Hydrological and Related Sciences

Because the importance of hydrological observatories has been recognized for decades, many existing field study sites are available that illustrate how innovative sensor technologies and modeling approaches are being, or could be, applied. In this chapter, case studies drawn from the expertise of committee members and their many collaborators are presented. They represent a broad range of different types of projects in terms of motivation, location, design, and duration of study. Some primarily describe on-going activities while others represent proposed activities that would be overlain on existing but limited monitoring or science programs.

Each example is presented to highlight one or more important issues related to the design and operation of hydrologic observatories, test beds, and campaigns. However, this report is not recommending that these specific case studies be undertaken. Further, the individual case studies are intended only to illustrate the kinds of sensors, sensor networks, or data analysis that could be valuable in integrated observation systems, rather than to provide specific advice to any governmental entity. Generalized findings and conclusions derived, in part, from the case studies taken as a whole are found in Chapter 5.

INTRODUCTION TO THE CASE STUDIES

"*Monitoring the Hydrology of the Everglades in South Florida*" provides an excellent example of a large, complex, integrated observatory designed for pressing water management needs in an ecologically sensitive area. This case study describes how the South Florida Water Management District (SFWMD) together with the U.S. Army Corps of Engineers and numerous other state, local, and tribal partners are working together on issues of water quantity and quality, flood management, and ecosystem protection. The case study describes a large monitoring network of hydrological, meteorological, and other sensors along with the hardware and software infrastructure needed for data collection and management. Some new sensor systems that potentially could be employed to enhance ecosystem monitoring are identified. The case study illustrates the importance of *interagency cooperation*, which can be crucial to the success of complex observatories for hydrological and related sciences.

"*Impacts of Agriculture on Water Resources: Tradeoffs between Water Quantity and Quality in the Southern High Plains*" addresses the impacts of agriculture on water resources, with a focus on semiarid regions where water availability is a critical issue and where cycling of salts has large-scale impacts on water quality. This type of study is particularly important given that world food needs will continue to increase, that many nations are turning to biofuels, and that climate change may worsen drought in many parts of the world. The case study discusses the types of measurement and monitoring programs that should be conducted to provide the necessary information to develop sustainable water and land resource management programs in the High Plains. In the High Plains, large time lags exist between forcing (land-use change) and response (increased recharge; change in water quality). Hence, the study provides a 'classic' example of why *long-term observations* often are critical, and why it is important that observatories have the capability to endure through changing budget cycles.

"*Hydrological Observations Networks for Multidisciplinary Analysis: Water and Malaria in Sub-Saharan Africa*" extends the case study examples to include the Developing World and direct issues of world health and medicine. This study demonstrates the importance of establishing consistency in sampling locations for different parameters, in this case analysis of climate and hydrological conditions versus malaria outbreaks. Proper coordination of physical, chemical, biological, and medical data collection, at appropriate spatial and temporal extents, is a key to inferring the controls on malarial outbreaks or the best methods for preventing such outbreaks. This case study thus emphasizes the axiom that the *nature of the research question or research hypothesis* plays an important role in the design of the associated observation network.

"*Achieving Predictive Capabilities in Arctic Land-Surface Hydrology*" explores a rudimentary strategy for robust remote sensing hydrology in the Pan-Arctic, to identify capabilities needed to link in-situ observations to satellite sensor-scale observations. The assertion in this case study is that the appropriate mechanism for achieving this linkage is through robust models that span the scales of the hydrologic processes. Given the difficulty of access to Arctic sites and the sensitivity of the Arctic ecosystems, there is a pressing need for autonomous sensing stations at point- through plot-scales, airborne platforms for plot- through watershed scales, and satellite remote sensing for sub-watershed through Pan-Arctic scales. This case study therefore highlights the need for collecting and integrating data at *different scales* and for developing novel methods for working in remote and challenging environments.

"*Integrating Hydroclimate Variability and Water Quality in the Neuse River (North Carolina, USA) Basin and Estuary*" focuses on the impact of human activity and hydroclimate variability on watershed nitrogen sources, cycling and export, and their impacts on fresh water and estuarine ecosystem health. As outlined in this case study, the problem requires a synthesis of hydrologic, ecosystem and anthropogenic water, carbon, and nutrient processes within a coupled watershed and receiving estuary. Nutrient management in the Neuse watershed focuses on nitrogen, which is the limiting nutrient in the estuarine system. New sensing methods are emerging that promise to transform our ability to measure and understand nitrogen ecosystem dynamics. Hence, this case study demonstrates the need for *integrating measurement of hydrologic, biogeochemical, and other ecosystem-related processes*, *and for building coordinated teams with interdisciplinary capabilities*.

"*Mountain Hydrology in the Western United States*" explains how snow in mountains of the West is the main source of the region's water, with downstream hydrologic processes (e.g., groundwater recharge) and interactions with ecosystems controlled by processes at higher elevations. Hence, it is critical to develop models for water and energy fluxes in the western mountains that can take into consideration not only past and present conditions but likely changes brought about by climate change. The critical issue is the need for high spatiotemporal resolution due to sharp wet-dry seasonal transitions; complex topographic and landscape patterns; steep gradients in temperature and precipitation with elevation; and high interannual variability. Given the need for high spatiotemporal resolution data, this study illustrates the need for developing and taking advantage of *emerging embedded sensor network* technologies, coupled with already existing monitoring and modeling strategies.

CASE STUDY I—MONITORING THE HYDROLOGY OF THE EVERGLADES IN SOUTH FLORIDA

The Everglades

The Florida Everglades (Figure 4-1) is one of the world's largest freshwater wetlands. It was once a free-flowing river of grass that provided clean water from Lake Okeechobee to Florida Bay. The marshes and swamps acted as natural filters that recharged underground aquifers in the South Florida region.
Historically the pre-channelized Everglades hydrologic balance was maintained through long, slow, continuous, gravity flow of water. Because of the diversion of water, channelization of transient rivers, and loss of elevation through oxidation of soils, pump stations are now required to move water from canals to marsh areas or from one canal segment to another or to return seepage water that would otherwise be lost from the greater Everglades. Over 50 such stations now exist, pumping volumes ranging from ~200 cfs to ~4800 cfs (Susan Sylvester, SFWMD, written commun., November 2006).

Accordingly, today the releases from Lake Okeechobee are controlled. During normal climatic conditions, Lake Okeechobee outflows are able to meet the large water needs to the south of the lake. However, when the climate remains abnormally dry for an extended period (for one or two seasons), inflows may diminish to very low levels during the same period that demands on the lake will peak. Consequently, lake stages may fall very quickly to extremely low levels. Conversely, when climatic conditions are wetter than normal, large volumes of water enter the lake, coinciding with periods when water demands to the south will be minimal. These events cause lake stages to rise very quickly and require large volumes of water to be discharged to the Water Conservation Areas (WCAs) or to the St. Lucie and Caloosahatchee estuaries. Abrupt changes in flow or very large releases through the estuaries are harmful to these ecosystems.

The WCAs are the primary source of supplemental water for the highly developed urban areas along the southeast coast of Florida, with the lake being the alternate source. The WCAs were built as large water-storage impoundments in the Everglades to provide both water supply and flood protection for the urban areas. In addition to the agricultural and municipal water consumptive needs, water releases from the lake are required to meet the needs of the Everglades and the numerous coastal ecosystems. The WCAs and the Everglades National Park (ENP) are known today as the remnant Everglades. Water held in and released from the WCAs effectively recharges the Biscayne aquifer in some areas.

Over the past half-century measures taken to satisfy agricultural and urban development goals have degraded the Everglades ecosystems. To restore and maintain the vitality of these ecosystems as well as to enhance the reliability and quantity and quality of water supplies, and provide flood protection, the U.S. Army Corps of Engineers (USACE), the South Florida Water Management District and numerous other federal, state, local, and tribal partners involved in water

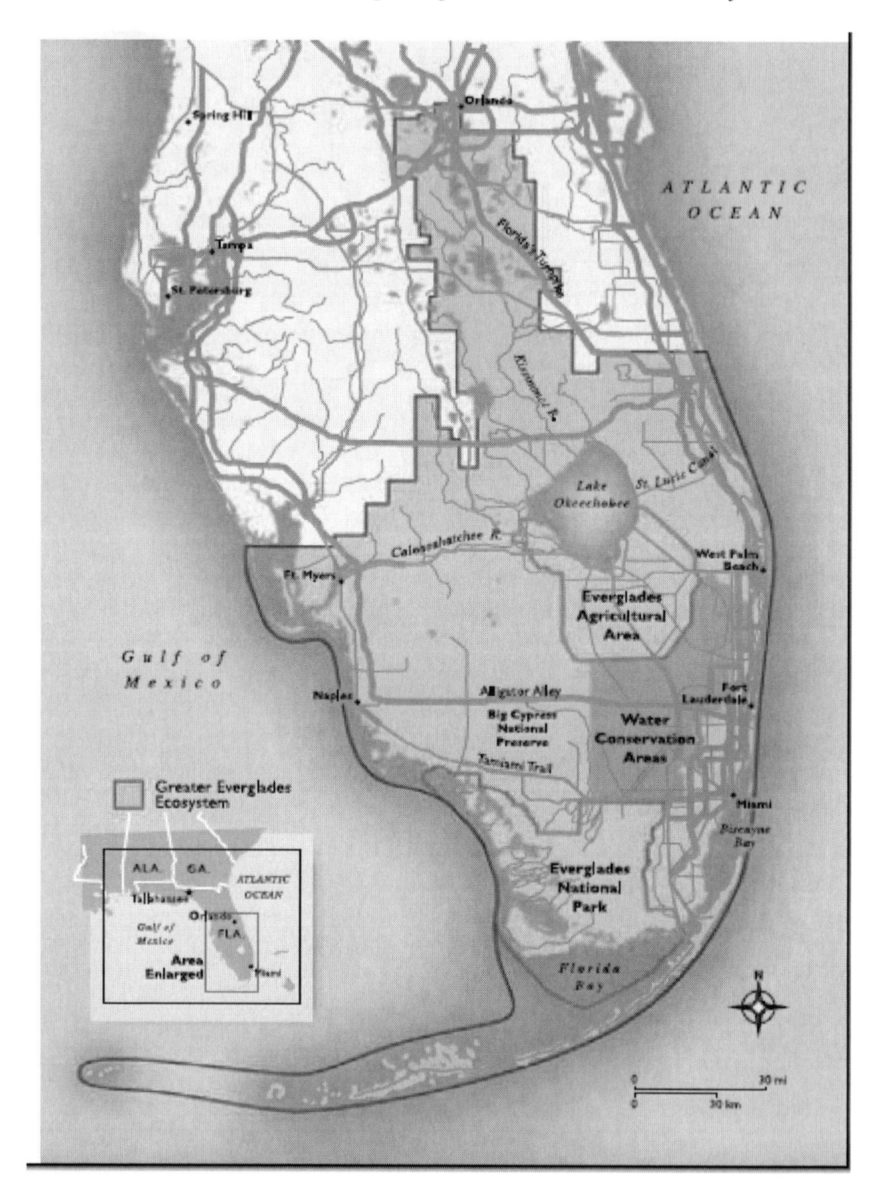

FIGURE 4-1 Map of the Everglades region in South Florida. The South Florida Water Management District is responsible for managing the hydrology and ecology in this area. SOURCE: NRC (2006a). © International Mapping Associates.

management in South Florida, have developed a plan called the Comprehensive Everglades Restoration Plan (CERP). To learn how to better manage the water, and to better understand the impact of various regimes of water-quantity flows, stages, volumes and qualities, and their durations and timing, timely, comprehensive, and accurate monitoring information is essential. Considerable sums of money have been spent in establishing an elaborate hydrologic, meteorological, and water-quality monitoring system throughout the Everglades. This is an excellent example of how an integrated hydrological observatory can provide essential data for managing water quality and quantity, flood control, and ecosystem protection.

The South Florida Water Management District

The South Florida Water Management District (SFWMD or the District) is the primary agency responsible for monitoring, managing, and protecting water resources in a 46,439 km (17,930 mi) region of South Florida. The District operates approximately 3000 km (1800 mi) of canals and more than 200 primary water control structures to serve a population of over 7 million people. The District's annual budget exceeds $1 billion of which some $20 million (about 2 percent) is spent on hydrologic monitoring and associated data management activities. (This number would be larger if such activities in the area of water quality were included.)

The hydrologic monitoring network of the District is divided into five parts: (1) rainfall, (2) meteorological, (3) surface-water stage, (4) surface-water flow, and (5) groundwater. These networks are spatially distributed over the geographic areas of the District. For each network, the District maintains records on the history and evolution of the network; information on sensor(s)/instrument(s) used; number and location of instruments; frequency of data collection; time interval of the available data; optimization or design of the network conducted; and relevant references used.

The District has been gathering data about the region's water and land resources for more than 40 years. Information about past and current weather, rainfall, and changes in vegetation or land use is essential for current and future planning, operations, research, and restoration initiatives. Real-time data, especially when combined with historic data, help the District make more informed water resources management decisions. Information about how natural and man-made systems are working (or not)—individually and interactively—is essential to short and long-term water resources management and restoration.

Data Collection and Management

Modern electronic hydrologic monitoring, data collection, and management began at the District in 1974. This has allowed water managers to remotely monitor strategic flood gates and control hydrologic conditions. The backbone of this system today is a 24-station microwave infrastructure with two-way radio extensions. This recently modernized microwave communications infrastructure now supports voice radio relay, supervisory control and data acquisition (SCADA), telephone circuits/trunks, and computer network traffic. SCADA systems include hardware and software components that scan all remote data, log data and system events, send alarms when abnormal conditions occur, and issue operator commands to remote devices. The District's SCADA and Hydro Data Management (SHDM) Department is responsible for data collection and management (Figure 4-2).

The District's SCADA system transmits and receives information on water stages or levels, wind velocities, rainfall, water temperature, salinity levels, and other data. The system operates continuously and uses wireless communications to monitor and control water level, water control gate positions, and pumping activities. It provides an early warning of possible flood problems by observing water level and rainfall trends. This computerized data collection system comprises the cornerstone of the District's data collection through a District-wide network of real-time and near-time data collection stations. The District also obtains and processes a variety of manual data logs.

Hydrologic data management includes processing the data collected, summarizing, deriving and analyzing, storing, and publishing. Processed data are archived into two different databases, namely, Data Collection/Validation Pre-Processing (DCVP) and DBHYDRO. Instantaneous (breakpoint) data are stored in the DCVP database, while daily summary and 15-minute interval data are published in the DBHYDRO database. End users can retrieve data from either of these two databases. DBHYDRO data are accessible to users through the web browser. Internal users can also retrieve information from the DCVP archive using the web browser.

The District maintains a structured quality assurance/quality control procedure to ensure that data collected is of the best possible quality before it is further published. Pre-processing is the first stage of operations applied to "raw" time series data collected within the District's monitoring network. Data records are collected and posted to DBHYDRO after data processing. Data quality assurance is normally performed during data processing. However, for some select legally mandated sites and for baseline data used in regional modeling and CERP, some post-processing quality assurance/quality control including graphical plotting and statistical analysis, are also performed.

Monitoring Networks

As of April 30, 2007, the District actively operates and maintains a network of 287 rain gauges to obtain rainfall data. These data are supplemented by radar

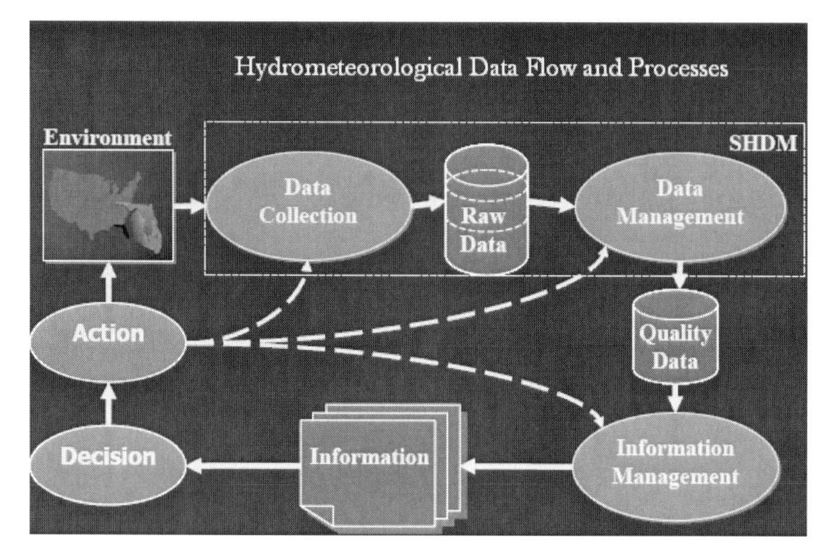

FIGURE 4-2 The hydrometeorological data flow and process. SOURCE: Pathak (2008).

rainfall NEXRAD (Next Generation Radar) data. The District also operates and maintains 45 active weather stations. In addition, data used to estimate daily potential evapotranspiration (PET) by "the Simple Method" several methods are available for 19 weather stations. A network of 1265 active surface-water stage gages provides surface-water stage data for various water bodies. Additionally, the District owns and operates a network of 446 active surface-water flow monitoring sites that provide instantaneous flow data at 15-minute intervals. From these data mean daily flows data are derived. The groundwater monitoring network has a total of 905 groundwater wells that are monitored on an interval basis of 15-minute, monthly, or greater than 1 month. The hydrologic monitoring network at the District is dynamic in nature and is constantly being expanded and optimized to the changing needs of the District.

Rainfall Measurements: The Importance of Employing Complementary Methods

An example of how different methods of data collection can complement one another is shown by the NEXRAD system. NEXRAD or Weather Surveillance Radar data provides complete spatial coverage of rainfall amounts unobtrusively using a predetermined grid resolution (usually 2 km × 2 km or 4 km × 4 km). The NEXRAD rainfall data is limited by reliance on the measurement of raindrop reflectivity, which can be affected by factors such as raindrop size and signal reflection by other objects. Because the reflected signal measured by the radar is proportional to the sum of the sixth power of the diameter of the raindrops in a given volume of atmosphere, small changes in the size of raindrops can have a dramatic effect on the radar's estimate of the rainfall. For this reason, the radar is generally scaled to match volume measured at the rain gages. The best of both measurement techniques is realized by using rain gage data to adjust NEXRAD values.

Surface-Water Flow Monitoring Network: Linking Monitoring with Control

The District operates a network of 446 active flow monitoring sites that are used in operations, planning, and regulatory aspects of water management. The flow monitoring network is shown in Figure 4-3. The District works closely with the U.S. Geological Survey (USGS), USACE, and various local agencies in measuring and/or estimating flow throughout the District's water control facilities. Water control structures are used to divert, restrict, stop, or otherwise manage the flow of water. These water control structures include pump stations, spillways, weirs, and culverts. District structures are typically designed to operate under a combination of water levels and operating regimes, which in turn result in varying flow conditions. Flow that moves through the structures are estimated by using a rating equation appropriate for the flow conditions based on the structure's static and dynamic data. The "static" data include the geometric characteristics of the structure, whereas the "dynamic" data comprise the water stages (headwater and tailwater) and operating conditions (gate opening and pump speed).

Groundwater Monitoring Network: An Example of Interagency Cooperation

Groundwater monitoring data are needed to assess long-term trends in groundwater availability; to develop, verify, and calibrate groundwater flow models; to assess temporal groundwater conditions during droughts; to provide data for water-use permit application evaluations; to assist the District in legal proceedings involving regulatory and other groundwater disputes; and to use in designing and evaluating various District projects. The District groundwater network consists of wells that have data publicly available through the District's DBHYDRO database but also in other databases not publicly accessible (these are mostly project specific). There are ongoing plans to migrate both of these databases into the DBHYDRO database. The groundwater network also con-

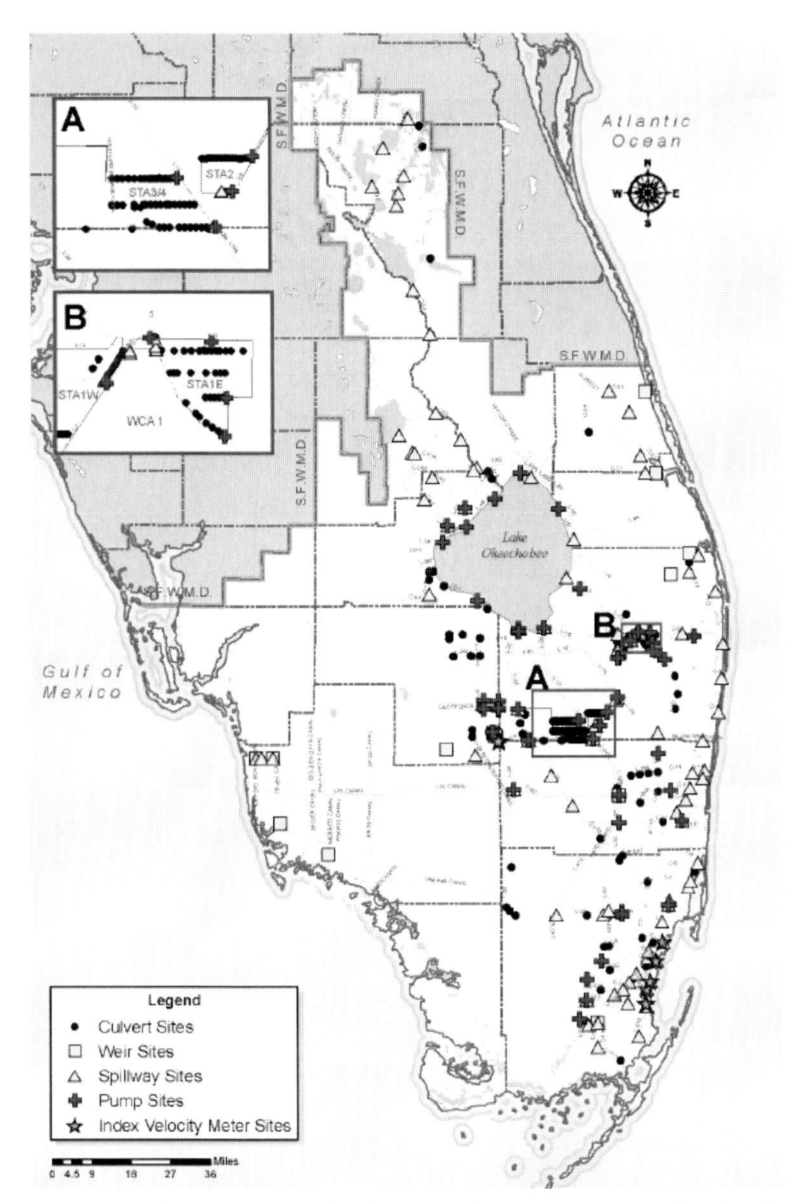

FIGURE 4-3 District flow monitoring network. SOURCE: Reprinted, with permission, from Pathak (2008).

sists of wells monitored by USGS through a cooperative agreement with the District. Most of the data on these wells are also available in DBHYDRO, but some can only be assessed from USGS's Automated Data Processing Systems (ADAPS) database.

The District measures groundwater levels by using a pressure transducer, typically connected to a data logger. The pressure transducers measure head pressure. The transducers communicate with the data loggers through an electronic cable. The data loggers then convert measured pressure values into water levels and record these data for subsequent downloads via laptop computers. Alternatively, data from some of the wells connected to the data loggers are sent via telemetry.

Ecological Monitoring: Merging Established and Emerging Approaches

The Everglades/Florida Bay landscape is a mosaic of different habitats that have evolved under a highly dynamic set of environmental conditions. As with any complex system, interactions among its different components are a fundamental aspect of its operation and play an important role in sustaining the Everglades. The physical hydrology, biogeochemical nutrient cycling, and biology of plant and animal communities are determinants of the emergent ecosystem properties that comprise the landscape. Monitoring these different "processes" that "drive" the system is providing data on how to best restore and maintain this dynamic landscape.

Monitoring the complex hydrologic, floral, and faunal changes associated with restoration activities is an enormous task. Many physical, chemical, and biological parameters have been identified as measures, or indicators, of overall performance of proposed restoration activities. Carefully designed and methodically implemented system-wide monitoring strategies are needed to successfully quantify both short- and long-term changes within the interdependent landforms, vegetation assemblages, and animal communities.

Due to the vastness (some 3 million acres) of the Greater Everglades, including the surrounding agricultural and urban environments, fixed monitoring stations (e.g., stage gages, water-quality collection sites), and vegetation/soil field sampling schemes (e.g., points, transects) cannot yield the high density of sampling data needed to adequately characterize and model the diverse ecosystems. Remotely sensed data, which is able to cover large areas with uniformly distributed high-density data points, has been and will continue to provide the essential synoptic view of restoration activities and their effects.

Work is underway to develop realistic and attainable strategies to expand the effective utilization of remotely sensed data for CERP system-wide adaptive monitoring and assessment. This involves the participation of biologists, chemists, hydrologists, and engineers in matching potential remote sensing technologies (sensors and analysis techniques) with the data required for effective water and ecosystem management. These data include vegetation types and patterns over time and space, including invasive species, and measures of habitat change and suitability for selected ecosystem indicators ranging from the ubiquitous cyanobacteria periphyton to the Florida panther. These data also include water quality (e.g., nutrients and mercury), vegetation, and food webs that support wading birds and tree island communities, changes in topography, and the nesting and feeding activities of alligators and various species of wading birds.

The Florida Everglades are ripe for integration of embedded sensor networks for hydrological and ecosystem monitoring with the already existing monitoring systems. Embedded network ecosystem monitoring systems are currently being developed at the University of California James San Jacinto Mountains Reserve biological field station (http://www.jamesreserve.edu; Hamilton et al., 2007). In addition to the biogeochemical and meteorological embedded sensor systems described in Chapter 2, researchers at the James Reserve are equipping bird nest boxes with microclimate sensors and a downward-pointing, miniature, infrared, sensitive charged-coupled device (CCD) camera and a diffused, infrared light-emitting diode (LED) light source. The cameras deliver JPEG images at 10 frames per second, which are reviewed by a staff biologist to record nest box activity during the nesting season. Such approaches could be used in the Florida Everglades to help monitor the health of species and ecosystems.

Summary

The fate of the Everglades is a dramatic case study of a global issue: freshwater quantification and allocation. Monitored data obtained from the District's databases provide water managers at any time with knowledge of the amounts of water available throughout the region. Based on this information, they make their flow allocation decisions that redistribute the water within the region each day.

Decisionmakers from around the world are watching South Florida, to see how wetland restoration will be balanced against economic development and societal demands. Scientists are currently developing interim goals and a strong monitoring and assessment plan for the Everglades. These efforts will supply the data needed to reduce the ecological and economic risks associated with adaptive restoration, and hopefully provide the framework for the successful restoration of this national treasure.

The Everglades serve as an example of several important axioms related to integrated observations of hydrologic and related sciences. First, integrated observatories can play an essential role for such critical activities as water management, flood control, and ecosystem monitoring. Second, observations in such politically, economically, and ecologically sensitive systems as the Everglades need to be done with great care, combining relatively simple and sophisticated and often expensive approaches. Finally, observational networks provide data, on which understanding can be built; however, understanding complex systems takes time.

Hence, as shown in the Everglades, there needs to be a long-term commitment to hydrological and ecosystem monitoring, with development and integration of new and emerging methods.

CASE STUDY II—IMPACTS OF AGRICULTURE ON WATER RESOURCES: TRADEOFFS BETWEEN WATER QUANTITY AND WATER QUALITY IN THE SOUTHERN HIGH PLAINS

Problem Statement and Importance

Agriculture has large-scale impacts on water resources. This case study focuses on semiarid regions where water availability is a critical issue and where cycling of salts has greatly affected water quality. Large reservoirs of salts (e.g., chloride, sulfate, and nitrate) have built up in soils as a result of evapotranspirative enrichment of atmospherically derived salts under natural ecosystems over millennia in many semiarid regions, such as Australia, the southwestern United States, and Africa (Figure 4-4) (Allison et al., 1990; Edmunds and Gaye, 1997; Phillips, 1994; Scanlon et al., 2007). This system is generally not dynamic and responds slowly to land-use and climate change. Changes in the water cycle related to land-use change can increase recharge and mobilize salt reservoirs, degrading groundwater and surface-water quality.

Conversion of natural ecosystems (grasslands or forests) to rain-fed agriculture often increases water resources through enhanced recharge (Figure 4-4). The classic example of increased water resources caused by cultivation is provided by conversion of forests to crops in Australia, which increased recharge up to two orders of magnitude, raising groundwater levels and baseflow to streams (Allison et al., 1990). In contrast, irrigated agriculture decreases water resources through elevated water use (Figure 4-4). Irrigated agriculture accounts for about 80 percent of global water withdrawal and about 90 percent of global water consumption (Shiklomanov, 2000). Irrigation is also one of the primary users of water in the United States and currently accounts for about 64 percent of freshwater withdrawal (excluding thermoelectric use) (Hutson et al., 2004). The water balance changes associated with irrigated agriculture cause changes in water quality. Increased recharge is generally associated with degradation of the quality of groundwater and surface water resulting from mobilization of salts in semiarid vadose zones and application of fertilizers in arid and humid settings (Figure 4-4). Reduced water quantity related to irrigation is generally associated with degradation of water quality also as a result of soil salinization and nutrient contamination from fertilizers.

In the future agriculture will come under increasing stress to meet the food requirements of projected increases in human population from 6 billion (2000) to 9 billion (2050). Many different approaches are being proposed for increasing food production, including expansion of rain-fed agriculture into nonagricultural areas (forests, grasslands), expansion of irrigated agriculture, increased use of fertilizers, and limited irrigation through rainwater harvesting in rain-fed areas. Development of sustainable agriculture in newly converted agricultural areas and remediation of water-quality and -quantity problems in existing agri-

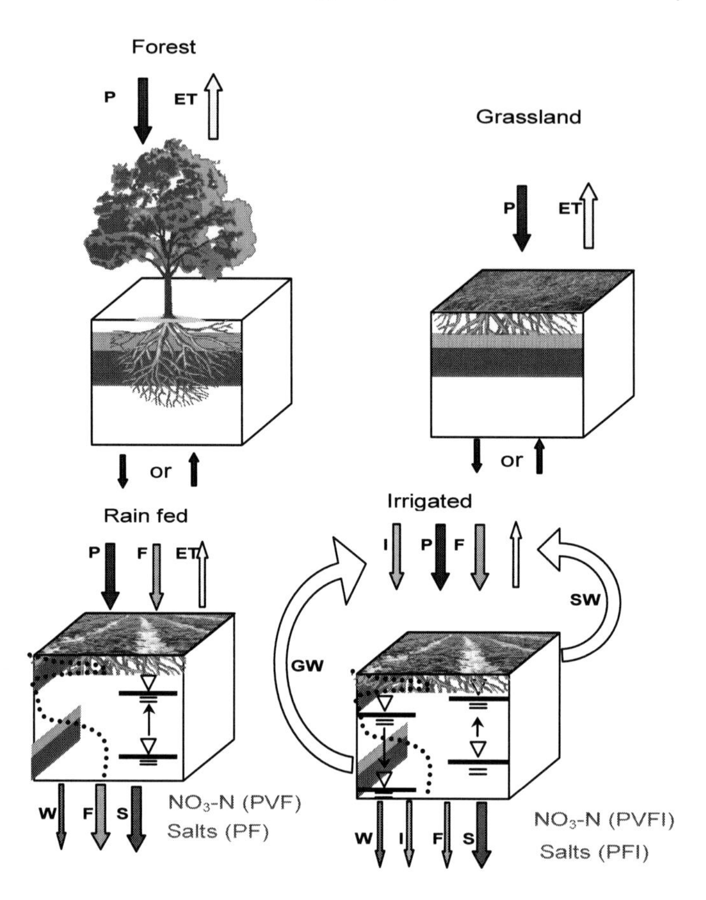

FIGURE 4-4 Schematic of different land-use settings: natural ecosystems including (a) forest and (b) grassland, agricultural ecosystems including (c) rain-fed and (d) irrigated agriculture. There is little or no recharge beneath natural ecosystems in semiarid and arid regions and salt and nutrient reservoirs are found in the unsaturated zone. Conversion of natural ecosystems to rain-fed agriculture results in decreased evapotranspiration and increased runoff (not shown) or recharge (W, water), raising water tables and mobilizing salts (S) in semiarid regions and nutrients from fertilizers (F) in semiarid or humid regions. Salts are derived from precipitation and fertilizers and nitrate from precipitation, N fixing vegetation (V), and fertilizers (F). Irrigated agriculture has an additional input of irrigation water that can be derived from surface water or groundwater. Irrigated agriculture generally results in increased recharge from irrigation return flow (I). Salts are mobilized by increased drainage in semiarid regions in groundwater-fed systems. SOURCE: Modified, with permission, from Scanlon et al. (2007). © 2007 American Geophysical Union.

cultural areas require a thorough understanding of how agriculture impacts water resources and land quality. Types of questions that need to be addressed include

1. How does rain-fed agriculture increase groundwater recharge?
2. How can water-quality problems caused by increased recharge and flushing of salts related to agriculture be remediated without greatly reducing water quantity?
3. Can natural ecosystems be converted to agriculture in the future with minimal impacts on water quality?
4. What impact do government programs that remove land from cultivation have on water resources?
5. What impact does conversion of agriculture to natural ecosystems (forests, grasslands) have on water resources (water quantity or quality)?
6. Can irrigation be managed to minimize negative impacts on water resources?
7. Can impacts of land-use change be reversed by simply reversing the land use?
8. What time lags are associated with forcing (land-use change) and response (impact on water or land resources)?

Many water resources problems related to agriculture are being addressed in different programs. Australia is trying to develop schemes to reverse dryland salinity problems caused by cultivation through crop management programs (reduced fallow periods, perennial versus annual crops) and afforestation without greatly reducing water quantity. In developing countries, rain-fed agriculture is projected to expand by 13 percent by 2030 to meet increased food demands of a growing population (Bruinsma, 2003). It will be important to manage such land-use changes to minimally impact water resources and land quality. In many areas, cultivated lands are being converted to natural ecosystems through unmanaged land abandonment, afforestation for timber production or carbon sequestration, and government policies such as the Conservation Reserve Program in the United States; however, the impacts of many of these land-use changes on water resources are generally not evaluated. Irrigated agriculture is projected to increase by 20 percent by 2030, further stressing scarce water resources (Bruinsma, 2003). There is increased interest in reducing water use associated with irrigated agriculture; however, extremely water-efficient irrigation systems can result in soil salinization because salts built up in the soils are not flushed through the system. The challenge will be to develop irrigated agricultural programs that are sustainable without negatively impacting land and water quality. Rotations of irrigated and rain-fed agriculture or minimal irrigation of rain-fed agriculture with rainwater (rainwater harvesting) may provide solutions to this problem (Rockstrom and Falkenmark, 2000).

The Southern High Plains

The southern High Plains provide an excellent case study for the problem of assessing impacts of agriculture on water resources (Figure 4-5). The basic issue addressed in this case study is which measurement and monitoring programs would need to be conducted to provide the necessary information to develop sustainable water and land resource management programs.

The High Plains (450,000 km^2 area), extending from South Dakota to Texas, are one of the most important agricultural areas in the United States. They represent 27 percent of the irrigated land and 20 percent of the groundwater used for irrigation in the United States (Dennehy, 2000; Qi et al., 2002). The southern High Plains (75,470 km^2) in Texas include 11 percent irrigated agriculture, 44 percent dryland or rain-fed agriculture, 32 percent grassland, and 12 percent shrubland (Scanlon et al., 2005). Although the percentage of land use represented by irrigated agriculture in the southern High Plains is low, irrigation accounts for about 94 percent of groundwater use and has resulted in groundwater-level declines (average 43 m over 10,000 km^2 over approximately 55 yr) from pre-development levels (Figure 4-6) (Qi et al., 2002; Fahlquist, 2003; McGuire, 2004). The southern High Plains are extremely flat, and most of the surface water drains internally to approximately 16,000 endorheic ponds or playas (Figure 4-5). Recharge is focused beneath playas; regional rates are about 11 mm/yr (Figure 4-7) (Wood and Sanford, 1995). There has been no recharge for the past 10,000 to 15,000 yr (since the Pleistocene) in natural ecosystem settings (grasslands and shrublands) in interplaya regions (Figure 4-7) (Scanlon and Goldsmith, 1997). Conversion of natural ecosystems to rain-fed, cultivated crops has increased recharge to about 25 mm/yr over a 3200-km^2 area (median value; about 5 percent of mean annual precipitation). Irrigated agriculture is fed entirely by groundwater and has resulted in groundwater depletion. The fate of drainage water or irrigation return flow is not known for most areas; however, studies by McMahon et al. (2006) indicate that irrigation return flow is still within the unsaturated zone and has not reached the water table. Groundwater quality is degrading, as shown by 31 to 221 percent increases in total dissolved solids and nitrate beneath cultivated areas (Scanlon et al., 2005).

Potential land-use and water-use scenarios that may be considered to maintain or increase food production and improve water resources include

1. Enhancing productivity of rain-fed agriculture,
2. Expanding agriculture into natural ecosystems (grasslands and shrublands), and
3. Increasing water- and nutrient-use efficiency of irrigated agriculture.

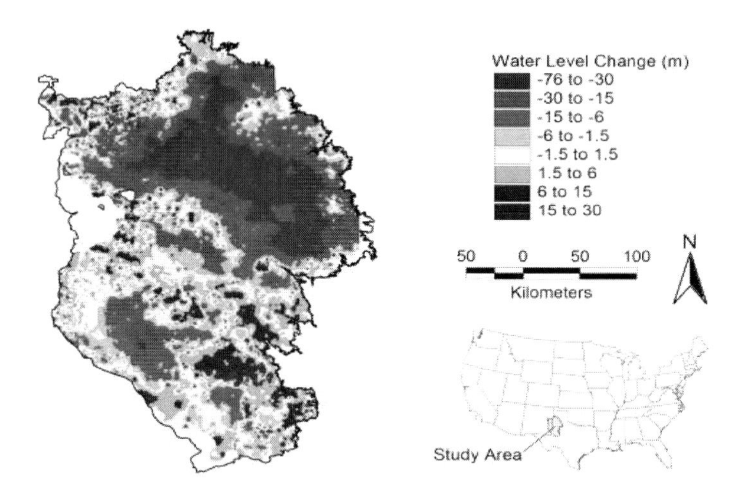

FIGURE 4-5 Groundwater-level changes caused in the southern High Plains, declines resulting from irrigation and increases resulting from conversion of rangeland to dryland agriculture. The area of greatest declines in the northeast region has an average value of 43 m (average pre-development time of 1945 to 2003). SOURCE: Modified, with permission, from Scanlon et al. (2005). © 2005 by Blackwell Publishing.

FIGURE 4-6 Typical view of a playa in the High Plains in Texas. There are approximately 30,000 playas in the southern High Plains that cover ~2 percent of the surface area. SOURCE: Photo courtesy of High Plains Underground Water Conservation District No. 1, Lubbock, Texas.

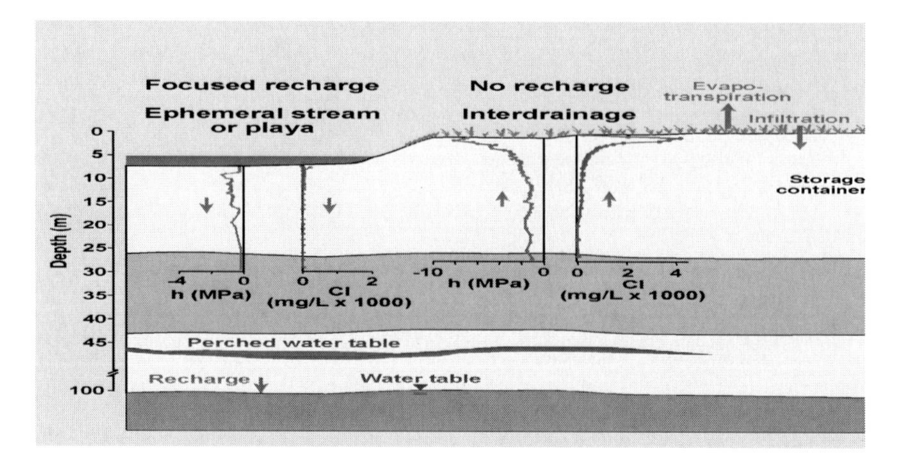

FIGURE 4-7 Schematic of water cycle in the southern High Plains, emphasizing focused recharge beneath playas and no recharge (evapotranspiration) in interplaya settings under natural ecosystems. Pressure heads close to 0 and low chloride concentrations in soil water beneath playas result from focused recharge. Upward pressure head gradients and bulge-shaped chloride accumulations beneath natural ecosystems in interplaya settings result from no recharge during the past ~10,000 yr since the Pleistocene glaciation. SOURCE: Based on data from Scanlon and Goldsmith (1997). © 1997 by American Geophysical Union.

Methods of enhancing food productivity of rain-fed agriculture include limited irrigation with rain (rainwater harvesting) over short dry periods, reduced fallow periods, terracing, even cropping to reduce bare soil evaporation and increase water use by crops (transpiration), and reduced drainage beneath crops (Scanlon et al., 2007). A large part of the southern High Plains (44 percent) is uncultivated. Can rain-fed agriculture expand into this region without negatively impacting water resources? Approaches to increasing water- and nutrient-use efficiency in irrigated areas include various irrigation systems, such as subsurface drip to reduce evaporation of irrigated water, precision agriculture using global positioning systems to collocate water and fertilizer applications with crop roots, various tillage approaches (no till, minimum tillage) to reduce runoff and increase infiltration, and perennial cropping to sequester nitrate and reduce groundwater contamination.

Current research in the southern High Plains is focusing on subsurface drip irrigation and water-quantity issues, with little emphasis on potential for soil and water salinization. Salinization may take decades to develop, and this issue may be more readily assessed using modeling analyses to determine the potential for salinization. Soil profiling at 5-year time intervals to measure water, salt, and nutrient balances and electrical conductivity changes over time associated with different types of irrigation systems would also be useful in tracking the development of salinization. Remote sensing of vegetation parameters, such as leaf area index, over long time periods may also be used to evaluate early development of salinization associated with deficit irrigation. Baseline data can be collected in areas with center pivot or other types of irrigation systems. Overall, such measurements and modeling analyses will be essential to assessing sustainability of different agricultural programs with respect to land and water resources.

Information Needs

Fundamental information on various fluxes and stores in the water, salt, and nutrient cycles would be required to understand potential impacts of different land-use changes related to agriculture. It is critical to consider linkages and feedbacks among these various cycles within the context of water and land resources. Fluxes include precipitation, irrigation application, salinity and nutrient application, evapotranspiration (ET), runoff, groundwater recharge, and groundwater pumpage. Storages that would need to be quantified include surface-water storage (ephemeral lakes or playas), soil moisture storage, groundwater storage, and salinity and nutrient storages in surface, unsaturated zone, and groundwater. At a minimum, information on these fluxes and stores would be required at seasonal timescales that can be scaled up to annual timescales. In addition, crop phenology (leaf area index, root growth) would need to be quantified.

Existing Capabilities

A network of precipitation gages monitors hourly and/or daily precipitation in the southern High Plains. Information on chloride, sulfate, and nitrate in precipitation is being collected annually as part of the National Atmospheric Deposition Program (two stations in southern High Plains, 1994-2004; http://nadp.sws.uiuc.edu). Information on nutrient application rates at the county level is based on county fertilizer sales data and is available from the National Agricultural Statistics Service (http://www.usda.gov/nass/). Solar radiation, required for estimating PET, was available from meteorological stations in Lubbock and Midland from 1961 to 1990; however, these stations were discontinued. A new PET network was established that includes weather stations with solar radiation for PET estimation (Texas South Plains ET network). Estimation of reference crop ET from potential ET requires information on crop coefficients. Crop coefficients are currently being quantified using weighing lysimeters at the U.S. Department of Agriculture (USDA) laboratory in Bushland in the central High

Plains (e.g., Howell et al., 2004). There is currently no monitoring of runoff to the 16,000 playas or variations in water levels in playas. Groundwater recharge has been estimated regionally using the chloride mass balance approach with groundwater chloride data (Wood and Sanford, 1995). Most of the recharge in natural rangeland settings is focused beneath playas. In addition, specific studies have been conducted at several sites throughout the southern High Plains to estimate recharge rates (Wood et al., 1997; Scanlon et al., 2005; McMahon et al., 2006). Regional recharge estimates have been provided by groundwater chloride data (Wood et al., 1997). Recharge beneath playas has been estimated using environmental tracers such as chloride and tritium (Wood and Sanford, 1995; Scanlon and Goldsmith, 1997). Recharge beneath rain-fed and irrigated agriculture has been estimated at about 10 sites using environmental tracers in the unsaturated zone and groundwater-level fluctuations in the saturated zone (Scanlon et al., 2005; McMahon et al., 2006). Groundwater levels have generally been measured once every 5 yr in the southern High Plains since the early 1900s, and data are available on the Texas Water Development Board (TWDB) website (http://www.twdb.state.tx.us). There is very little information on groundwater pumpage. Recently a program was implemented to install meters in irrigation wells to begin to characterize this flux.

There is little information on surface-water storage. Water levels in playas are not being monitored. Limited monitoring of soil moisture is being conducted at a few sites in natural ecosystems and irrigated sites (Scanlon et al., 2005). This system was installed as part of the USGS National Water Quality Assessment Program in the High Plains. Soil moisture is monitored annually by USDA prior to crop planting using a neutron probe. Groundwater storage can be estimated from annual groundwater level measurements conducted by the TWDB and is available on-line. Salinity and nutrient storage information is not available for surface water. Limited information on salinity and nutrient storages are available for the vadose zone in natural ecosystems and rain-fed and irrigated sites funded by USGS National Water-Quality Assessment (NAWQA) Program and Bureau of Reclamation (Scanlon et al., 2005; McMahon et al., 2006). This information is extremely valuable for quantifying potential loading of salts and nutrients from the vadose zone to groundwater caused by increased recharge related to cultivation. Long-term records on groundwater concentrations of major and minor elements are available for the past 40 to 60 yr on-line at the TWDB website. This excellent database of very high quality, which is provided by the state, is not available in most states within the United States.

A detailed groundwater model has been developed for the southern High Plains. All the data used to develop the model are available on-line at the TWDB database (http://wiid.twdb.state.tx.us).

Information Gaps

There are various gaps in our measurement and monitoring system of the water, salt, and nutrient cycles that need to be addressed. Many of the questions related to agriculture posed earlier can be evaluated using baseline measurements of environmental tracers, salts, and nutrients in the unsaturated zone. Before natural ecosystems are converted to agriculture, it is important to characterize the stores of salts and nutrients in the vadose zone because increased recharge related to cultivation can mobilize these salts and contaminate underlying aquifers, as seen in Australia (Allison et al., 1990). The issue is more critical for nitrate contamination relative to contamination with sodium chloride salts because of the lower maximum contaminant level (MCL) for nitrate-N (10 mg/L) relative to total dissolved solids (TDS) or chloride (500 mg/L TDS, secondary MCL for Cl, 250 mg/L).

Time series in groundwater quality over decades beneath agricultural areas would be required to assess impacts of agriculture on groundwater quality. The time lag between conversion to agriculture and increased groundwater recharge can also be evaluated using chloride profiles in the unsaturated zone in agricultural areas. Velocity of the solute front movement related to cultivation can be estimated beneath rain-fed agriculture and projected in order to estimate the time that the increased flux reaches the water table. Time lags are commonly long (decades to centuries). Therefore, negative impacts of land-use change may not be immediately apparent. In addition, attempts to reverse negative impacts of land-use change, such as dryland salinity in Australia, will also require long time periods.

A combination of unsaturated zone and groundwater data for nitrate can be used to determine whether fertilizers are being leached beneath cultivated areas. However, most measurements related to agriculture are restricted to the root zone, and rarely are profiles measured below the root zone or to the depth of the aquifer to assess leaching of nutrients. Chloride profiles can also be measured beneath irrigated agriculture to determine whether deficit irrigation, as is practiced in the southern High Plains, results in soil salinization. Preliminary results from limited profiles in the southern High Plains indicate that soil salinization is an issue because chloride concentrations in irrigation water are relatively high (≥500 mg/L TDS) and recharge from precipitation is insufficient to prevent this buildup of salts. These data suggest that current trends toward more efficient irrigation systems (e.g., ≤95 percent efficient) will increase salinity of the applied water by a factor ≤20. By focusing on water quantity alone, these programs are ignoring impacts on land resources.

Large time lags between forcing (land-use change) and response (increased recharge) often result in limited use of monitoring to address many of these issues. However, monitoring can provide process understanding related to impacts of cultivation on water and nutrient cycles. Monitoring ET over different land uses, such as natural ecosystems and rain-fed agriculture, can be used to determine whether reduced ET associated with winter fallow periods in agricultural areas can account for increased recharge beneath rain-fed agriculture. If ET monitoring proves that fallow periods lead to increased recharge, then programs to reduce fallow periods to increase food production would reduce groundwater recharge. Impacts of reversing land-use changes to remediate water-quality problems associated with cultivation or abandoning agricultural areas can also be examined using ET data.

The current system of using potential ET and crop coefficients is based on limited spatial distribution of meteorological stations and crop coefficient data. This information could be greatly enhanced by using optical satellite-based data (Landsat Thematic Mapper [TM], Moderate Resolution Imaging Spectroradiometer [MODIS]) in programs such as Surface Energy Balance Algorithm for Land (SEBAL) and Disaggregated Atmosphere-Land Exchange Inverse Model (DisAlexi) to estimate regional ET (Bastiaanssen et al., 1998a, b; Anderson et al., 2004). One of the main problems with the satellite data is that the thermal band on Landsat TM is no longer functioning, and if the National Aeronautics and Space Administration (NASA) does not reinstitute a high-resolution thermal imaging system on future Landsat platforms for regional ET estimation this problem will continue indefinitely. The spatial scales of the Landsat TM (30 m) are appropriate for estimating ET at the field scale; however, data are available only once every 17 days. In contrast, the resolution of temperature data from MODIS is much coarser (250 m), although the temporal resolution is twice daily. Both Landsat and MODIS data can be used together to offset differing spatial and temporal resolutions.

Recently developed Large Aperture Scintillometer (LAS) systems can be used to provide information on sensible heat flux at scales up to 1 km. The relatively large scale of these LAS measurements makes these data suitable for ground-referencing sensible heat-flux estimates from satellite-based approaches. In addition, LAS systems can be used to estimate ET by monitoring other components of the energy balance equation (net radiation and soil heat flux). ET can also be estimated using GRACE (Gravity Recovery and Climate Experiment) satellite data (Rodell et al., 2004), using a water balance approach. The large footprint of GRACE data indicates that reasonable estimates of ET may be possible only for the entire High Plains system because of coarse spatial resolution. The proposed GRACE follow-on mission would likely improve the spatial resolution considerably.

Although monitoring of surface runoff is generally routine, very little information is available on spatial and temporal variability in storage in endorheic ponds or playas. Satellite data can be used to monitor variations in the surface area (Smith et al., 2005) and water depth (Alsdorf and Lettenmaier, 2003) of these playas; however, ground-based information on water levels and topography of playa floors would be required to translate area or depth information to storage volumes. Recent reductions in irrigation over the past two decades and projected future reductions in irrigation may result in reduced irrigation return flow to playas, which would negatively impact these habitats for migrating birds. Monitoring playa water storage is also very important for recharge estimation because playas are the focal points of recharge (Scanlon and Goldsmith, 1997). Although many studies document impacts of climate variability on surface runoff, studies examining impacts of climate variability on endorheic water storage are limited.

Estimates of groundwater recharge are needed for all the different land-use settings in the region, which can be partly addressed through unsaturated zone profiles using environmental tracers such as chloride (Scanlon et al., 2005; McMahon et al., 2006). Soil physics monitoring can also be used to determine flow directions (upward or downward). If these recharge estimates can be linked to surface parameters such as vegetation or leaf area index, there is a potential to use satellite-based vegetation indicators to estimate spatial variability in recharge.

Groundwater storage is currently being estimated from annual groundwater-level monitoring data. GRACE satellites can also provide estimates of monthly, seasonal, and annual changes in groundwater storage over the entire High Plains (Rodell et al., 2002, 2006; Strassberg et al., 2007), as well as for smaller, regional aquifer systems (Yeh et al., 2006). Recent post-processing studies indicate that GRACE is capable of providing reliable information on water storage variations as much higher resolution (~150,000 km^2) than was previously thought possible (Swenson et al., 2006). However, monitoring of groundwater levels should be increased to seasonal or monthly timescales to provide ground validation of the GRACE output. A recently developed commercial field superconducting gravimeter can also be used to estimate changes in subsurface water storage (precision <0.1 microgal, equivalent to water-layer thickness of 24 mm) over spatial scales similar to water-table depths (Peter et al., 1995; Goodkind, 1999). Survey gravimeters can be used to extend results from the point gravimeter to larger spatial scales.

Modeling analyses can be used to integrate measurement and monitoring data from different space and timescales, isolate controls on flow and transport processes, and conduct sensitivity analyses with respect to proposed future land-use changes and potential impacts on water and nutrient cycles. Because of generally long time lags associated with impacts and responses related to initial land-use changes or remediation schemes in these semiarid regions, it is essential to use modeling to assess different proposed management strategies. A variety of codes are available to simulate appropriate processes in this type of system, including land atmosphere, regional water balance, unsaturated zone, and groundwater codes.

Measuring and monitoring various components of the water cycle, such as precipitation, ET, runoff, and groundwater recharge, associated with different land-use practices can be used to provide information to water resources managers on linkages between land-use and water resources, including both water-quantity and water-quality issues. A comprehensive evaluation of total system performance that includes land use and degradation, water quantity, and water quality is essential for considering trade-offs between water-quantity and water-quality impacts of different land-use changes, as well as land degradation associated with salinization. Although traditional approaches focus on single agricultural programs such as rain-fed agriculture or irrigated agriculture, combining or rotating these approaches may help to minimize negative environmental impacts. Rotating rain-fed agriculture with irrigated agriculture may help minimize salt buildup beneath deficit irrigated areas. Limited irrigation of rain-fed agriculture is being proposed in sub-Saharan Africa to increase food productivity of rain-fed systems and may be appropriate in the southern High Plains also. Modeling analyses can be used to evaluate more complex rotations. Results of the measurement, monitoring, and modeling analysis in the southern High Plains may be applicable to many water-scarce, semiarid regions facing similar problems of increasing food production with more limited water supplies, such as in Australia, Africa, India, and China.

Summary

The High Plains case study emphasizes the importance of linking land-use and water resources management and also the tradeoffs between water-quantity and water-quality impacts of land-use changes. Long time lags between land-use changes and water resource impacts, such as changes in recharge and salinization, place increased emphasis on use of environmental tracers such as chloride that archive impacts of land-use changes on groundwater recharge and salinity. Monitoring programs that quantify the various components of the water cycle and integrate ground-based and satellite information on precipitation, ET, runoff, and groundwater storage, can provide process information to assess controls on land-use change impacts on water resources. Sustainable water resources management in semiarid regions, such as the southern High Plains, will require an in-depth understanding of land-use change impacts on water resources and will likely require a shift from irrigation to rain-fed agriculture in many areas.

CASE STUDY III—HYDROLOGICAL OBSERVATIONS NETWORKS FOR MULTIDISCIPLINARY ANALYSIS: WATER AND MALARIA IN SUB-SAHARAN AFRICA

Malaria: A Global Health Problem

The nature of the research question or research hypothesis should play an important role in the design of the associated hydrological observation network. If the research question requires integration of knowledge and data from several disciplines, then it is necessary to design a network of observations that includes a suitable set of sensors or sensor systems, not necessarily limited to the traditional set of sensors used in hydrologic research. The observations network should be designed to sufficiently characterize the relevant physical, biological, and/or chemical processes at the appropriate spatial and temporal resolutions, and with sufficient precision and accuracy to be able to make useful decisions. In this section, a case study on the challenges of monitoring the interactions between the hydrology and the ecology of malaria in Africa is presented.

This is only one example of a broad set of water-related diseases. In the United States, mosquito-borne disease transmission (e.g., West Nile virus) is an important public health issue. Addressing these important research problems requires integration of concepts and tools from climatology, hydrology, entomology, and medicine. The approach used here is an example of landscape epidemiology, which involves identifying areas where a disease is present and/or transmitted. Key environmental factors such as temperature, precipitation, elevation, vegetation, and water levels, are combined with human factors such as population centers, land use, and transportation networks, and with characteristics of the disease itself (range, mode of transmission, etc.). Numerous examples of this approach are given at http://geo.arc.nasa.gov/sge/ health/landepi.html.

Malaria profoundly affects people around the world, killing 3 million out of 300-500 million infected each year. The majority of the victims are children under the age of five (Persidis, 2000). This alarming rate has waned in the past as anti-malarial drugs became available but is now in a period of rapid resurgence in sub-Saharan Africa. It is estimated that in 1995, 200 million Africans were infected with malaria, and of these 1 million died (Snow et al., 1999). This recent trend has been widely reported (Hay et al., 2002), and as a result the disease has once again become a priority. Classic malaria drugs such as quinine and chloroquine have become substantially less effective as pathogens have developed immunity to them (Persidis, 2000), a factor that is likely to be partly responsible for the rising malaria death rates. Much of the ongoing research seeks to develop new drugs, vaccines, and other prevention strategies.

The societal concerns about malaria are not limited to conditions under the current climate. The International Panel on Climate Change (IPCC) concluded that malaria will likely be on the rise as a result of climate change. However, investigators disagree about the extent of the effect of climate change on the rate and geographical extent of malaria incidence (Hay et al., 2002). Previous attempts to link the two processes have focused on statistical analysis of climate variables and malaria data at nearby locations. These attempts lack consistency in the spatiotemporal resolution and locations of the two different types of data, which are derived from observations networks in hydrology and human health that were designed independent of each other. In order to address this limitation, an observations network has to be designed in a consistent fashion to sample all relevant variables. The distribution of vectors for the mosquitoes carrying the disease depends on local climatic factors, and the occurrence of outbreaks and outbreak intensity are likely to depend on local weather and hydrology (Epstein et al., 1998). These connections are evident in Figure 4-8 from Niger, which illustrates clearly how reported incidence of malaria peaks following the rainy season.

Towards Better Understanding of the Connections between Water and Disease

A group of scientists from the Massachusetts Institute of Technology's School of Engineering, the Harvard School of Public Health, the National Oceanographic and Atmospheric administration, and the Pasteur Institute has recently initiated a multidisciplinary research project that brings together their complementary expertise to address the complex interactions that lead to malaria transmission. This project is motivated by the following set of research questions:

1. What is the role of local environmental conditions and microclimatic niches in dictating the local dynamics of individual mosquitoes and mosquito populations (*dynamics* being defined as mean level of abundance and rates of biting and survival)? What physical and biological processes dominate the complex interactions that take place between mosquito populations, human populations, and the environment?
2. What are the relative roles of natural and man-made surface-water bodies in providing suitable habitat for sub-adult mosquitoes?
3. Can the response of mosquito populations to seasonal and interannual climate variability in Africa be predicted?

The aim of this project is to develop an integrated computational model that is capable of simulating the spatial and temporal dynamics of malaria transmission in sub-Saharan Africa. Such a model is likely to provide a new tool for prediction of malaria outbreaks and in the screening of different strategies for combating malaria transmission, and will provide fundamental data on climate associations with malarial outbreaks that will enable assessment of the impact of climate change on malaria transmission and its regional extent. A detailed nu-

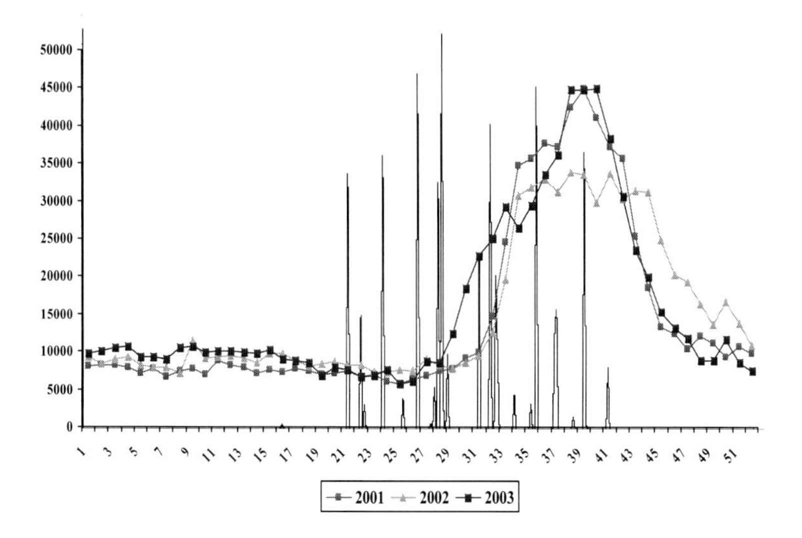

FIGURE 4-8 Weekly malaria cases in Niger from 2001 to 2003 and precipitation data (average monsoon conditions).

merical representation of all the physical and biological processes that govern the rate and extent of malaria transmission is included. The researchers envision a numerical simulator that will explicitly describe, at fine spatial and temporal resolutions (~10 m, 10 minutes), the interactions between the natural environment and the populations of mosquitoes, parasites, and humans. Such a tool will provide a suitable framework for integration of hydrological and entomological observations with other information about soil type, vegetation cover, and topography.

The three main objectives of the project are

1. To build consistent data sets on climatic conditions, hydrologic conditions, mosquito populations, and malaria incidence; these data sets are needed for characterization of processes, calibration of model parameters, and verification of model performance;
2. To develop a discrete model of mosquito population dynamics incorporating variables affecting the development rate, pupal productivity, and adult longevity of *Anopheles gambiae* s.l. mosquitoes; and
3. To couple this mosquito population model to a detailed, fine-scale, hydrologic model that describes the natural environment for mosquito population including seasonality, distribution, and persistence of suitable larval habitats.

In order to achieve objective 1, the foundation for the other objectives, a network was designed to collect observations of rainfall, incident solar radiation, air temperature, humidity, wind speed and direction, soil saturation, microtopography, levels in water pools, spatial extent of pools, groundwater level, mosquito density, larva count, and malaria incidence (see Figure 4-9).

The design of this network integrates sensors that are traditionally used in hydrology for monitoring rainfall rate and soil moisture level, and in entomology for monitoring mosquito density and larva count, as well as standard meteorological sensors of temperature, humidity, and wind. The monitoring of the micrometeorology and soil hydrology is automated, and these observations are recorded by data loggers at about hourly resolutions. The weekly monitoring of the larva prevalence is done by dipping small containers into the water body of interest followed by visual identification, and counting. The monitoring of adult mosquito density is carried using standard U.S. Centers for Disease Control (CDC) light traps followed by laboratory identification of the mosquito species, and counting. There are needs for development of new automated sensors for monitoring the mosquito population at the adult and larva stages. Automatic cameras can be deployed to monitor the larva prevalence in water bodies. New sensors that respond to the movement and associated sound of adult mosquito may provide information about their density.

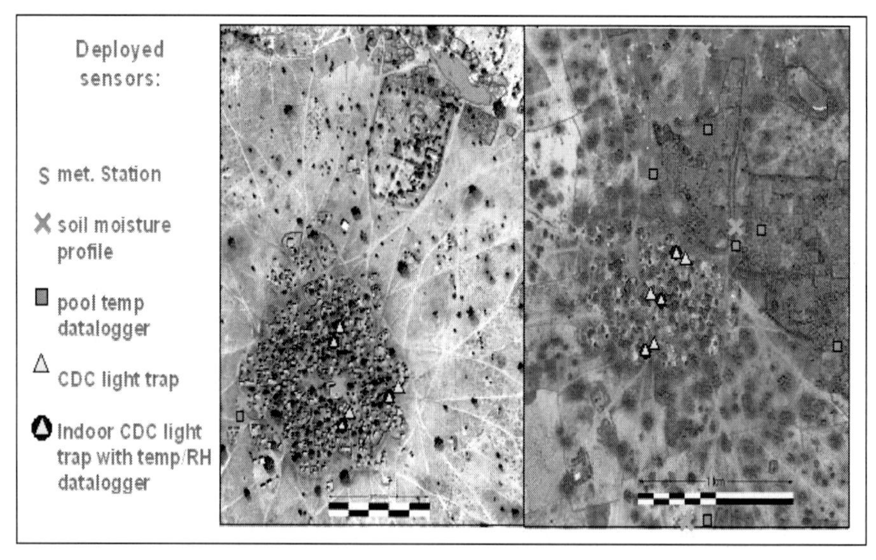

FIGURE 4-9 Deployed sensors in villages of Banizombou and Zindaru, near Niamey in Niger. CDC - U.S. Centers for Disease Control, RH - is relative humidity.

The density of the sensors in Figure 4-9 characterizes the spatial resolutions resolved by the network. The resources needed for the labor-intensive processes of field counting of larva and laboratory identification of adult mosquito species are the main limiting factors for expanding the density of the deployed sensors. Extrapolation of these observations from the village scale to the regional scale would require the use of remote sensing technology, from satellites and other airborne platforms, in order to study the same processes at the regional scale. The new generation of active microwave sensors (such as those deployed in RADARSAT2) will offer a great potential for monitoring of water bodies that are likely to provide the breeding sites for mosquito. These sensors will enable monitoring of these water bodies at a resolution of a few meters.

A New Role for Remote Sensing Observations in Global Health Research

Because ground observations of surface-water prevalence are time consuming and difficult to carry out over large areas, an attractive alternative has been to use remote sensing measurements of land-surface wetness. Many such top-down studies have associated the abundance of vectors or vector-borne disease incidence with satellite imaging (e.g., Rogers and Randolph, 1991; Hay et al., 1996). Such investigations have generally used vegetation classification or the Normalized Difference Vegetation Index (NDVI), which measures vegetation greenness, as proxies for soil moisture and land-surface wetness. However, the NDVI and other similar indices were designed for monitoring vegetation conditions in large-scale climate studies, and do not capture all the detailed environmental conditions that influence the processes leading to malaria transmission. For addressing the malaria problem, there is need for developing new indices that characterize, among other variables, the fraction of the large area covered by persistent water pools that is suitable as habitat for sub-adult mosquitoes. In order to develop these indices one needs to use remote sensing data (such as products from active microwave remote sensing) that describe the size of the land area covered by water pools. The formation of water pools is affected not only by meteorological conditions such as rainfall and evaporation rates but also land-surface conditions such as topography and soil type.

Key elements to make significant progress in this important research area, as well as other similar areas of multidisciplinary research, would include the following:

1. Design the integrated observations network to suit the specific research problem at hand, instead of using observations that were collected with the objective of addressing a different research problem; for this project, the hydrologic and climatological sampling are being coordinated with the sampling of mosquito populations and health data so that the spatial and temporal sampling strategies will be comparable.

2. Identify the appropriate spatial and temporal resolutions that are needed for characterizing and modeling the interactions within the environment that are critical to answering the research question; this involves sensitivity testing of the model in combination with field study.

3. Integrate information about the environment from several types of measurements including ground, airborne, and spaceborne measurements. Such integration is best achieved through optimal assimilation of data into appropriate environmental models (see Chapter 3).

4. Integrate embedded sensors with community-based observations and remote sensing by using database management technologies.

5. Integrate information on the physical, biological, and chemical environments using a consistent multidisciplinary framework.

6. Develop a "collaboratory"—a web-based system of data, predictive models, and management projects for access by researchers and users in different fields and with e-mail and chat options for cross-disciplinary discussion.

7. Incorporate the recent advances in sensor technology, computing technology, geographic information systems, and satellite remote sensing.

Summary

This case study has shown an application of integrated observations to a major global disease that kills millions and infects hundreds of millions each year. This study demonstrates the importance of establishing consistency in the spatiotemporal resolution and locations of four very distinct types of data—in this case data related to climatic conditions, hydrologic conditions, mosquito populations, and malaria incidence. These data exist in part, but they come from observations networks that were designed independent of each other. Without coordination of physical, chemical, biological, and medical data collection at appropriate spatial and temporal extents, and data assimilation and modeling to move across scales and simulate the dynamics of malarial transmission, it would be difficult to infer the controls on malarial outbreaks or the best methods for preventing them.

CASE STUDY IV—ACHIEVING PREDICTIVE CAPABILITIES IN ARCTIC LAND-SURFACE HYDROLOGY

Climate change associated with global warming is arguably occurring most dramatically in the Pan-Arctic because, to first order, the forcing of the Arctic water cycle is categorically governed by the phase transition between ice and liquid water. The Pan-Arctic, an area defined by the Arctic Ocean and the lands that drain into the Arctic Ocean, encompasses a significant portion of the Earth's land area. How the Arctic climate evolves will influence the planetary heat balance and the circulation of the global oceans. Rapid climate warming, expressed as drying soils, rising average temperatures, and melting permafrost, is changing the ecosystem (Olsson et al., 2003), shortening winter access for oil and gas extraction (NRC, 2003), and damaging roads and structures on the tundra (Nelson et al., 2002). Two consequences, transition of tundra as a carbon sink to its being a carbon source (Oechel et al., 2000) and decreasing albedo (Chapin et al., 2005), serve as positive feedbacks to climate warming. Current warming is already having an impact upon the indigenous people of the North. The melting of permafrost presents challenges to infrastructure, rising sea levels are forcing abandonment of some coastal villages, and some traditional hunting and fishing practices are no longer viable (USGCRP, 2003).

The importance of the Arctic water cycle to global processes argues that these processes be well monitored as a means of checking the predictive accuracy of our models and, as humankind begins to manage the global climate, as a means to monitor the effects of our actions. There are currently temporal gaps associated with the difficulty of access to many regions of the Arctic in winter, spatial gaps associated with national willingness or ability to maintain comprehensive monitoring systems, and thematic gaps associated with the emerging comprehension of the impact of climate change in the Arctic. The need for an integrated hydrologic monitoring system in the Pan-Arctic prompted creation of a series of recent research programs. Among these are the international Arctic Monitoring and Assessment Programme (AMAP), the National Science Foundation (NSF) Arctic Research Consortium of the U.S. (ARCUS) Study of Environmental Arctic Change (SEARCH), and, most relevant to this report, NSF's Pan-Arctic Community-wide Hydrological Analysis and Monitoring Program (Arctic-CHAMP) (Vörösmarty et al., 2001). The overarching theme of these reports is captured in the National Research Council report, *Toward an Integrated Arctic Observing Network* (AON) (NRC, 2006b). An AON would integrate all relevant physical, biological, and social observing and data management elements to comprise an Earth observing system for the Arctic—essentially the Arctic component of the Global Earth Observing System of Systems (GEOSS). An integrated Arctic land-surface hydrologic observing system would be a subset of the AON.

Both the Arctic-CHAMP and AON proposals are comprehensive and compelling. Both identify remote sensing as an integral element of any Arctic observing system. However, neither differentiates among the potential contributions of airborne and satellite sensors, nor identifies the research and supporting infrastructure needed to fully realize the potential of satellite remote sensing. The objective of this case study is to explore a rudimentary strategy for robust remote sensing hydrology in the Pan-Arctic, specifically, to identify capabilities needed to meaningfully link in-situ observations to satellite sensor-scale observations. The assertion here, and as a general assertion of this report, is that the appropriate mechanism for achieving this linkage is through robust models that span the scales of the hydrologic processes.

The report from the workshop defining Arctic-CHAMP observed, "The water cycle of the Arctic plays a central role in regulating both the planetary heat balance and circulation of the global oceans. Recent and unprecedented environmental changes, such as declines in the total area of winter snow cover on land and declining sea ice cover throughout the Arctic Ocean, are now well documented. The causes of these changes and their impact on the global ocean and atmosphere are still poorly understood. The cycle of freshwater in the Arctic land atmosphere-ocean system is central to these observed changes" (Vörösmarty et al., 2001). This same report lists several key unresolved issues, namely:

- What are the major features (i.e., stocks and fluxes) of the Pan-Arctic water balance, and how do they vary over time and space?
- What are the hydrologic cycle feedbacks to the oceans and atmosphere in the face of natural variability and global change?

Among the principal recommendations of the report are

- A substantial commitment should be made to rescue, maintain, and expand current meteorological and hydrological data collection efforts, and
- An assessment of the feedback mechanisms through which progressive hydrological change influences both the natural and human systems is urgently needed (Vörösmarty et al., 2001).

That is, an observing infrastructure is needed that offers more information about the current Arctic hydrological system at all scales, and captures key aspects of the system's evolution with climate warming.

Any design of an Arctic hydrologic observing system is constrained by the temporal and spatial scales of processes being observed, and by the difficulty of access in winter, because of weather, and in summer, because of fragility of the active layer. Temporal scales of interest vary from hours, e.g., significant precipitation, to days, e.g., spring melting of the snow pack, to seasons, e.g., melting and refreezing of the active layer. The spatial scales of interest vary from point-scale for intrinsic measurements like temperature or soil moisture, to plot-scale for measurements like moisture fluxes in the active layer, to catchment-scale to relate drainage to topography, to watershed-scale for issues of water balance, to Pan-

Arctic-scales for the global water cycle. An example of an almost integrated Arctic hydrologic observing system would be the sum of the building blocks identified in the AON report. These are the Toolik Lake LTER (point- and plot-scale measurements in three characteristic terrains of Arctic tundra, Figure 4-10), the Imnavait Creek microscale system (a 2.2 km^2 headwater catchment), and the Kuparuk River watershed. Each has been extensively instrumented and studied.

Difficult access in winter or summer argues for autonomous sensing stations at point through plot scales, embedded sensor networks for plot through catchment scales, airborne platforms for plot through watershed scales, and satellite remote sensing for catchment through Pan-Arctic scales (Figure 4-11). Elements comprising proven autonomous stations, though not cheap, are commercially available for many key measurements. Satellite instruments have grown in number and sophistication. The weak elements are mature technologies for embedded sensor networks and ready access to airborne sensing technologies. These systems, were they available, overlap the scales of autonomous sensing at the pot-scale end of the scale spectrum and satellite-scale sensing at the other end of the scale. If those robust models that link hydrologic processes at plot-scale to the watershed-scale are to be developed and validated, embedded sensor networks, airborne sensors, and the interpretive skills to integrate the observations with in-situ data and satellite data are essential.

Embedded Sensor Networks

Given the fragility of the active layer and the difficult access, Arctic areas would benefit greatly from the new and emerging sensor technologies described in Chapter 2. Use of in-situ microcosms for measurement of chemical weathering rates has been demonstrated in the Antarctic (Maurice et al., 2002); a combination of static and continuous flow in-situ reactors coupled with temperature sensors and piezometers could be highly beneficial for studying hydrobiogeochemical processes in the active layer of soils and hyporheic zones during the warmer months. Microelectrode sensors, such as those used by Haack and Warren (2003) (Figure 2-3), would be useful for probing biogeochemical processes in tundra soils. Wireless embedded sensor technologies would have the potential to expand tremendously the spatiotemporal resolution of sampling, particularly of parameters such as incident sunlight (including various ultraviolet and visible wavelengths and photosynthetically active radiation) and temperature, both of which are cheap, easy to measure using current technologies, and can be integrated into tiny wireless devices. The mobile networked infomechanical systems (NIMS) tram cable sensor systems for surface-water, soil, and associated microatmospheric sampling shown in Figure 2-8 (Harmon et al., 2006) would be particularly useful to prevent damage to fragile soils and plant communities in locations where stable columns could be safely and reliably emplaced.

Dry Heath (Shrub) near Toolik Lake LTER. June 2005. Commonly found on windblown ridgetops.

Tussock Tundra near Toolik Lake LTER. June 2005. Commonly found on mesic (i.e., neither wet nor dry) hillslopes.

Wet Sedge near Toolik Lake LTER. June 2005. Commonly found in low-lying, waterlogged areas (note the standing water).

FIGURE 4-10 Three terrain types found on the Alaskan North Slope. These and the open water of thermokarst lakes comprise the "canonical" terrain types that might be used to characterize the low-relief regions of Arctic tundra. SOURCE: Reprinted, with permission, from England and De Roo (2006), University of Michigan.

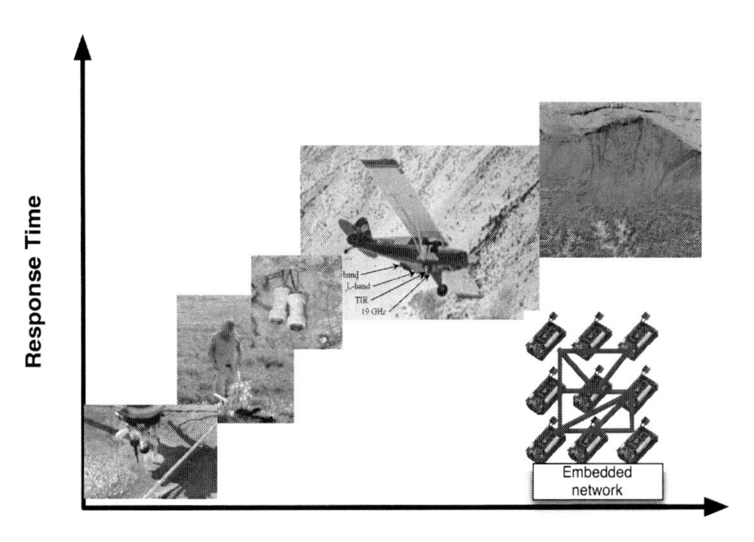

FIGURE 4-11 Schematic showing the geographical coverage versus response time of different sensing approaches as discussed in the text. The x axis shows how spatial extent of sampling increases from microsensors and in-situ sensors through airborne sensors to satellites, while the y axis demonstrates that the increase in spatial extent can come with a cost (in terms of actual cost and/or in terms of amount of time between design of a system, implementation, and return of data to investigators). A sensor network or sensor web allows increased spatial sampling with less time for deployment and data return, because the sensors are meant to be easily deployable and widely distributed with continuous data feed to web-based platforms. SOURCE: False color satellite image from NASA; microsensors courtesy of L. Warren, McMaster University, Canada; in-situ sensors courtesy of E. Boyd and G. Geesey, Montana State University.

Considering the sensitive environment, remote locations, and difficult terrain, a variety of challenges will need to be considered and overcome, such as:

1. difficulties with battery life and solar recharge systems in cold climates;
2. need to make systems capable of resisting severe winter conditions, including freeze-thaw, or to emplace and remove seasonally;
3. environmental considerations; if sensors are widely scattered, for example dropped from airplanes, then the sensors themselves could become a new source of widespread environmental pollution.

Nevertheless, such remote and difficult terrains are some of the prime locations whereby wireless sensor technologies could have the greatest effects on data collection. Key to this will be further development of robust systems and of new types of sensors for a wider variety of biogeochemical parameters.

An Example of a Potential Airborne System for Observing Moisture in the Active Layer

Radiometers operating below the Debye relaxation frequency of liquid water (~10 GHz) are sensitive to moisture in the upper few centimeters of soil (Jackson et al., 1984, 1995; Schmugge and O'Neill, 1986; Jackson and O'Neill, 1987; Schmugge and Jackson, 1992; Njoku and Entekhabi, 1996). Radiometers at 1.4, 6.9, 19, and 37 GHz, among other frequencies, have long been carried on large NASA aircraft to support terrestrial hydrology. A similar remote sensing approach is appropriate for the Arctic, but new digital technologies (Fischman and England, 1999; Fischman et al., 2002; Pham et al., 2005) enable fabrication of compact systems that can be carried by low-cost, Piper Cub-class aircraft. These aircraft can operate with very little maintenance out of unimproved landing strips, which would allow them to support season-long campaigns in the Arctic.

Airborne observations would allow investigations of the scale-dependent distribution of soil moisture that span the spatial gap between point and satellite observations. For example, Advanced Microwave Scanning Radiometer – EOS (AMSR-E), a microwave radiometer currently flying on NASA's Aqua satellite, produces data that are sensitive to moisture in the Arctic tundra. Figure 4-12 shows plot-scale, microwave brightness observations of tussock tundra during a diurnal cycle for several of the AMSR-E frequencies. Diminished brightness at 6.7 GHz relative to the higher frequencies is a consequence of moisture in the active layer. While this moisture sensitivity is expected, there have not been investigations to quantify the interpretation nor to extend them to the scale of the satellite footprint, ~50 km. Of potentially greater relevance for soil moisture sensing, the European Space Agency Soil Moisture Ocean Salinity (SMOS) mission, being developed for launch in 2007, will be the first satellite to carry a 1.4 GHz imaging radiometer (Kerr et al., 2001). Like AMSR-E, its footprint will be ~50 km. Without investigations of remote sensing signatures of scale-dependent processes between point observations and the SMOS spatial resolution, SMOS data are likely to have limited impact upon our understanding of land-surface hydrology in the Arctic.

An airborne sensor system capable of spanning the point-to-satellite scales is illustrated in the design study shown in Figure 4-13. Digital radiometer technologies enable robust designs of this power, weight, and size, but none have been built for the aircraft environment (Fischman and England, 1999), and no organization has the appropriate airborne platform to carry them. Airborne systems, like that in the design study shown in Figure 4-13, could be used to cali-

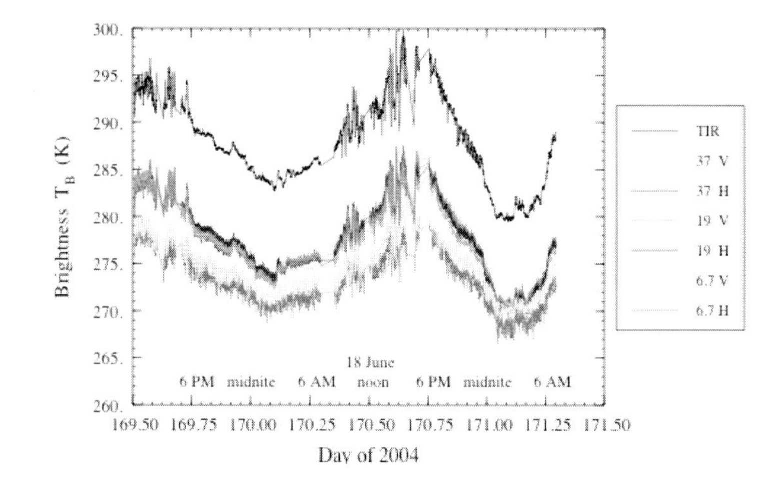

FIGURE 4-12 Diurnal temperatures at a 35^0 incidence angle for 6.7, 19, and 37 GHz brightness and for the thermal infrared radiometer (TIR) for tussock tundra. Lower brightness values for the lower frequencies are caused by moisture in the active layer of the tundra—clearly an important parameter for evaluating climate change and carbon cycling in such terrain. SOURCE: Reprinted, with permission, England and De Roo (2006), University of Michigan.

FIGURE 4-13 An Aviat Husky aircraft showing location of radiometers in a pod attached to the float fittings of the aircraft. The design study included 1.4, 6.9, and 19 GHz nadir-viewing, profiling radiometers, and a thermal infrared (TIR) profiling sensor. All sensors have a 17° beamwidth yielding 30 m footprints at 100 m heights above ground. Radiometers were Dicke-capable with two-point calibration and digital detection. 60 kn airspeeds yield 110 ms integration times with a 10 percent footprint smear, a NEΔT of ~0.8 K at 1.4 GHz, and ~0.4 K at higher frequencies. Modified, with permission, from Aviat Aircraft, Inc., Afton, Wyoming.

brate and validate land-surface models of the energy and moisture fluxes that are being developed for the canonical terrains shown in Figure 4-10. Their data would also help disaggregate the inherently low spatial resolution satellite data for the canonical terrain types within the radiometer footprint. Such systems enable hydrologic process studies at spatial scales that are difficult to capture with point measurements. Relatively inexpensive systems like this address two of the three uses of airborne systems identified in Chapter 2. They enable interpretation of satellite data, and contribute to the science of land-surface hydrology as it occurs in the Arctic. Both are essential elements of understanding and skillfully predicting the rapid evolution of the Arctic climate.

Summary

The Pan-Arctic region is experiencing rapid climate change associated with global warming, expressed as drying soils, rising average temperatures, and melting permafrost. There is a compelling need for an integrated hydrologic monitoring system in this region. This case history has explored some opportunities for hydrologic remote sensing in the Pan-Arctic, with a specific example given for observing moisture in the active layer of the soils. Opportunities range from microsensors and embedded sensor networks through airborne and satellite observations. Given the difficulty of access to Arctic sites and the sensitivity of the Arctic ecosystems, there is a pressing need for these technologies. However, there are also enormous challenges to developing and maintaining in-situ sensors and sensor networks for use under such extreme conditions.

CASE STUDY V—INTEGRATING HYDROCLIMATE VARIABILITY AND WATER QUALITY IN THE NEUSE RIVER (NORTH CAROLINA, USA) BASIN AND ESTUARY

Problem Statement and Background Information

This case study focuses on the impact of human activity and hydroclimate variability on nitrogen sources, cycling and export in coastal watersheds, and their impacts on fresh water and estuarine ecosystem health. The problem requires a synthetic treatment of hydrologic, ecosystem, and anthropogenic water, carbon, and nutrient (WCN) processes within a coupled watershed and receiving estuary. The focus was on the Neuse River Basin and Estuary (NRBE) in North Carolina. Nutrient management in the NRBE focuses on nitrogen as the limiting nutrient in the estuarine system, although interactions of nitrogen, phosphorus, and other constituents require an integrated, multicomponent approach.

Over the last two decades, major debates have developed over the condition and management of the NRBE, involving water quality, quantity, agriculture, urban sprawl, and coastal development. Headlines in local newspapers were filled with coverage of massive fishkills in the Neuse and its estuary, harmful algae blooms, the development of anoxic dead zones in the estuary, and their links to watershed runoff and nutrient loading from rapidly expanding industrial hog operations, intensified row crop production, urban sprawl, and renewed hurricane activity. Burkholder et al. (1997) reported a newly discovered dinoflagellate with a complex life cycle including a toxic and predatory stage stimulated by nutrients and other substances as a cause of fishkills and a toxic threat to human health. The popular press referred to *Pfisteria piscicida* as "flesh eating", because of the open lesions found on many of the dead fish. Reports of adverse human health responses to direct contact with estuarine water or inhalation of toxins emerged in the Neuse and as far north as tributaries of the Chesapeake. The complexity of the linkages between water quality, anoxia, microbial dynamics and fishkills, uncertainty in causal mechanisms, and general lack of adequate monitoring led to scientific and management controversy. Significant disagreement and debate in scientific publications and the popular press ensued, and an emergent response from state and municipal government, nongovernmental organizations (NGO), and other industrial and community groups developed within a framework characterized by strong positions, but very little observational data. One major source of agreement was that nitrogen loading from the watershed was a prime cause of the lower river and estuarine eutrophication. Since then, significant additional monitoring and sampling has been developed in the estuary, but an integrated sampling plan sufficient to characterize and explain watershed WCN dynamics is yet to develop.

The Neuse is a ~15,000 km^2 watershed draining into the Neuse Estuary, part of the Pamlico-Albemarle Sound complex (Figure 4-14). Nitrogen from point

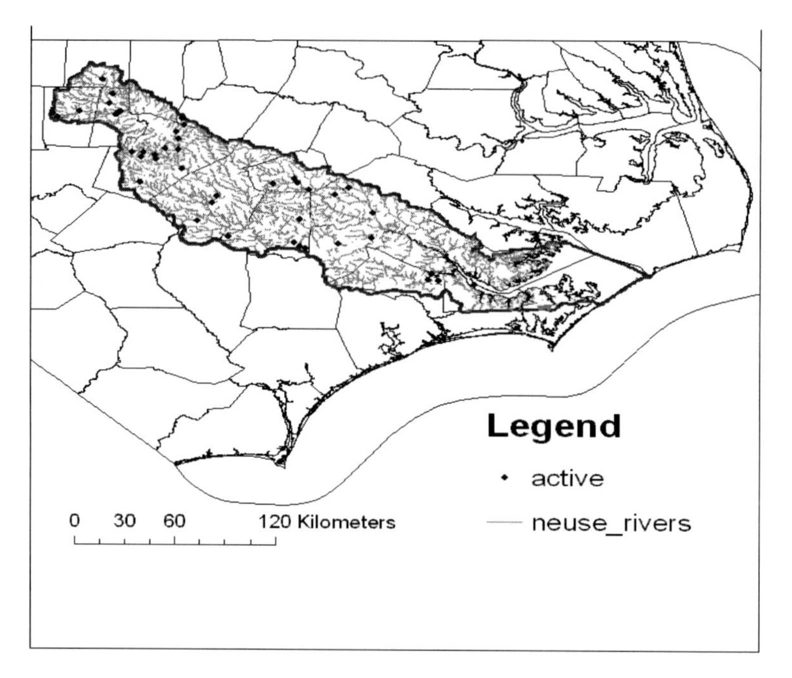

FIGURE 4-14 Location of currently active stream gages in the Neuse watershed, North Carolina.

and non-point sources cycles through multiple media and flowpaths in the atmosphere, surface water, soils, and groundwater as well as the built agricultural, residential, and industrial environment. Complex coupling between the stores and fluxes of WCN and variable hydroclimatic forcing (with record drought and floods in the last decade) create episodic and spatially heterogeneous loading, transformations and export of nitrogen from the terrestrial to the aquatic system. Nitrogen (N) flushing, referring to a build-up of terrestrial nitrogen stores during dry periods and subsequent mobilization during storm events, provides nitrogen pulses on top of chronic loading. Hydrologic state, through soil moisture influence on redox conditions and microbiological activity, is an important determinant of biogeochemical transformations of N. As an example, nitrification and denitrification, the microbially mediated transformation of nitrogen promoted by different redox conditions as influenced by soil saturation levels. Riparian zones, with typically organic-rich, moist soils, are often sites of high denitrification potential, converting nitrate into N_2 or N_2O gas. However, during dry periods, riparian soils may become aerobic and transform from largely denitrifying to nitrifying conditions, becoming a source of nitrate. Therefore, sources and sinks of runoff and nutrients can be highly variable in space and time, corresponding to "hot spot" and "hot moment" phenomena (McClain et al., 2003).

River discharge-driven variations in estuarine residence time and nitrogen flushing have been associated with spatial and temporal patterns of estuarine phytoplankton assemblages, repeated algae blooms, low dissolved oxygen events, and trophic system disturbance (Paerl et al., 2006a,b; Burkholder et al., 2004). Efforts to decrease nitrogen entering the NRBE require reductions in N point and non-point sources, as well as increasing ecosystem uptake and retention by best management practices (BMPs). These BMPs include land-use zoning, buffer strips, detention basins, constructed wetlands, stream restoration, and a set of other practices. Non-point sources are difficult to quantify as they are spatially diffuse and episodic, including large confined animal feed operations (CAFO), row crop and residential land use, and atmospheric deposition. A major impediment to the design and implementation of N reductions has been a general lack of understanding of the following:

1. what the individual and cumulative effects of BMPs are, particularly their performance over the distribution of hydrologic events the area is subject to,
2. where and when the bulk of N loading and delivery to the river and estuary occurs,
3. where and when N retention occurs, including storage and transformation into forms that can be later released, as well as denitrification and volatilization loss, and
4. what the magnitude and timing of river and estuarine ecosystem response would be to specific reductions in nitrogen loading.

The issues described in this section have immediate, practical applications. Regulation of N loading into the NRBE and implementation of BMPs are priorities for municipal, state, and federal agencies charged with protecting fresh water and estuarine ecosystems and resources. The issues and questions are also at the core of the field of ecohydrology, particularly the interaction of hydrologic storage and flux of water with ecosystem material (carbon and nutrient) balances. Progress in these areas are limited by a general lack of modeling frameworks that couple WCN interactions at the watershed scale, as well as observational programs with a sampling regime appropriate to the space and timescales characterizing key processes and feedbacks.

To date, the collection and synthesis of hydrologic and ecosystem information has not been coordinated as part of a centralized informatics system, as contrasted to the South Florida Water District. A set of monitoring programs in the NRBE funded by state and federal agencies are building important data sets on specific aspects of watershed and estuary ecosystems but requires enhanced observational methods, modeling, and institutional support and organization. Progress has been made on integration of diverse information stores through the state coordinated Modeling and Monitoring (MODMON) Project (http://www.unc.edu/ims/neuse/modmon), a prototype (paper) design of a Hydrologic Observatory (Reckhow et al., 2004). A more integrated project leveraging on these efforts could take advantage of recent advances in instrumentation, informatics, and modeling to provide a more fundamental understanding of the controls on magnitude and timing of N export, and response of the estuary. Such a project would need to last for a few to several years, in order to sample a range of hydroclimatic conditions with sufficient replication of seasonal dynamics to determine sources, sinks, and variability in the system sufficient to develop and test management options. Longer-term monitoring at reduced intensity would be coupled with distributed models and available remote sensing resources to track and evaluate temporal trends of N cycling and transport in response to management activities, land use, and climate change.

Sampling Strategy

Given the size and heterogeneity of the NRBE, nested sampling is designed to generate ecohydrologic storage and flux information to be scaled to the full watershed with the use of remotely sensed and other spatially distributed information and ecohydrologic models. The nesting would proceed at increasing levels of spatial detail from the full watershed to selected subwatersheds covering dominant land use and geomorphology, and hillslope/riparian systems within the subcatchments. Information required over this scale range includes

1. Land-use specific runoff quantity, pathways, and nutrient loads in wet-dry conditions,
2. Residence time, retention, and transformation of N in terrestrial and aquatic systems, and
3. Interaction of watershed discharge and N loading with estuarine dynamics.

Existing Measurement Capability and Information Synthesis

Currently available instrumentation and informatics in the Neuse are briefly described, and improvements for more efficient use of current technology, as well as new measurement technologies, are proposed following the discussion in Chapter 2. The existing major monitoring networks within the NRBE, and how they could be augmented to address the major research and management questions, are also described.

Much of the large-scale spatial information describing the Neuse (available on-line at http://www.env.duke.edu/cares/neuse/GIS.html) has been incorporated into the Neuse digital watershed (Maidment et al., 2005; Merwade et al., 2005). Figure 4-15 shows the locations of a series of these data layers for different measurements (or centroids for Next Generation Radar [NEXRAD]), each

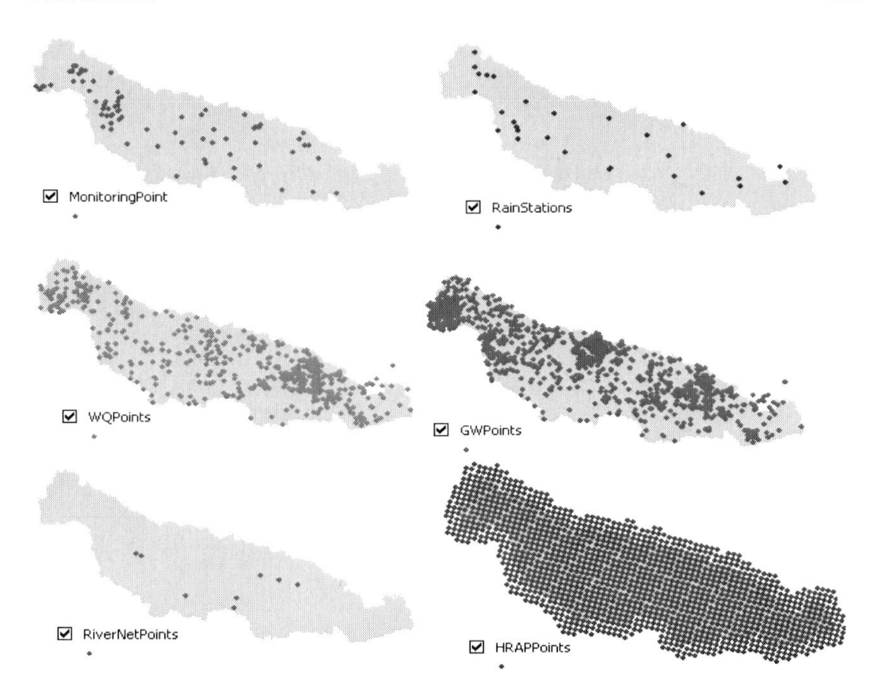

FIGURE 4-15 Point locations associated with time series objects within an Arc-Hydro digital watershed representation of the Neuse. MonitoringPoints are USGS NWIS streamflow stations, RainStations are National Climatic Data Center (NCDC) rainfall stations, WQPoints are National Water Information System (NWIS) water quality measurement stations, GWPoints are NWIS groundwater wells, RiverNetPoints are water quality stations operated by North Carolina State University (http://rivernet.ncsu.edu), and HRAPPoints contain center points of the Hydrologic Rainfall Analysis Project (HRAP) grid for NEXRAD precipitation estimates. SOURCE: Reprinted, with permission, from Merwade et al. (2005). © 2005 by CUAHSI.

linked to a time series of data values. Additional point feature classes are associated with centroids of the North American Regional Reanalysis (NARR) grid cells, and are linked to time series of 3 hr and monthly simulated meteorological, energy balance, and soil moisture values for the NARR time domain of 1979-2003. Additional information provided by a group of federal and state agencies and universities includes

1. **Surface-water discharge:** ~40 active USGS stream gauges and 5-10 additional gages operated by universities.

2. Surface-water quality*:* Sampling of inorganic and organic stream nutrients, pathogens, sediment load, and other water-quality parameters carried out at USGS gages and other state and local sampling sites, typically at quarterly to monthly frequencies.

3. The Rivernet Program (http://www.rivernet.ncsu.edu): Sampling of water quality at high frequency (1 hr) of water temperature, conductance, pH, NO_3, and turbidity on six higher order streams.

4. Soil moisture/temperature: Rooting depth (0.2 m) soil moisture and temperature are measured at several sites by the North Carolina State Climate Office (http://www.nc-climate.ncsu.edu/cronos/map).

5. Groundwater measurements: USGS groundwater measurements are typically limited to one measurement at the time of well completion, concentrated in the Coastal Plain. More intensive groundwater monitoring networks include a site with ~66 wells installed and monitored by USGS and North Carolina State Department of Water Quality in a CAFO in the Contentnea Creek subwatershed to study transport and transformation of hog waste-derived nutrients in complex Coastal Plain groundwater systems (Spruill et al., 2004, Tesoriero et al., 2005) and drought monitoring wells.

6. Atmospheric inputs: National Weather Service (NWS) NEXRAD rain radar systems in Raleigh and Wilmington provide full coverage for the NRBE. The North Carolina State Climate Office operates and distributes meteorological information that report 15-minute to daily precipitation, and min/max temperature collected from a network of different measurement sites (http://www.nc-climate.ncsu.edu) with a subset of sites providing hourly measurements of temperature, relative humidity, precipitation, solar radiation, and photosynthetically active radiation. Dry and wet N deposition are measured and estimated at National Atmospheric Deposition Program and Clean Air Status and Trends Network sites in and around the NRBE.

7. Estuarine physical, chemical, and biological state: Estuarine monitoring is carried out by state- and federal-funded efforts, including fixed buoy, periodic ship transects (http://www.marine.unc.edu/neuse/modmon) and a novel method of high spatial and temporal frequency measurement carried out in association with the North Carolina Department of Transportation (http://www.ferrymon.org). Physical, chemical, and biological information on the state and dynamics of the lower river and estuary are regularly collected, geolocated and distributed on these sites, and integrated into hydraulic and biogeochemical modeling of the estuary (http://www.unc.edu/ims/neuse/modmon).

8. Remote sensing information: High-resolution remote sensing products are available that characterize land cover, topography, and estuarine conditions. These include

1. A high-resolution land-cover classification using recent Enhanced Thematic Mapper, System Pour l'observation de la Terre, and aerial photographic interpretation that has been developed by the U.S. Environmental Protection Agency (Lunetta et al, 2003).

2. Annual land-cover change products from Moderate Resolution Imaging Spectroradiometer (MODIS) imagery (http://maps6.epa.gov).

3. LiDAR elevation data produced by the state of North Carolina, the Federal Emergency Management Agency, and other agencies include statewide bare earth (BE), 20' and 50' elevation grids (http://www.ncfloodmaps.com).

4. Remote sensing of estuarine and lower river water quality carried out by high spectral resolution airborne imagery (AVIRIS, CASI) detection of chlorophyll and phytoplankton assemblage (Figure 4-16) for selected time periods.

Information Gaps—Extension of Existing Resources

To adequately address the research questions posed above, additional measurement and monitoring of new variables, and extension of current sampling regimes at key scales and locations need to be planned. Reckhow et al. (2004) designed a prototype Hydrologic Observatory sampling scheme for the Neuse supplementing existing monitoring programs, over a scale range that includes major geomorphic zones and land-use conditions in the watershed. The elements of a hypothetical extended network are summarized in Table 4-1. Readily available instrumentation, as well as emerging and envisioned measurement technologies that we expect may be available over the next decade, are both included. The overall goal is to locate individual sensors as part of a multiscale and multidisciplinary observation network. Plot-scale sampling will be clustered within gaged subcatchments, reflecting the theoretical paradigm of the hillslope ecohydrologic models that are used as both a conceptual framework to study N flushing mechanisms and modeling in this case study. Sampling from ridge to riparian zone would capture hydrologic and biogeochemical evolution along hillslope surface and subsurface flowpaths within the drainage areas of small gaged catchments. Sampling along specific channel reaches would investigate in-stream routing and nutrient processing of water from the contributing catchments. In the Coastal Plain, sampling would be arranged along flow gradients accounting for potential lack of correspondence between surface and subsurface flow directions. One gap that would be filled is the addition of small, land-use-specific catchments. The existing network would be expanded following the methods suggested by (Maidment, 2004) using a GIS-driven allocation to develop both nesting and representativeness of the gage network for a desired range of land use and drainage areas. As an example, Figure 4-17 shows percent urban land use in subcatchments in the upper NRBE extracted with a 10-km^2 drainage area threshold from the North Carolina LidAR dataset and the land-use classification of Lunetta et al. (2003).

The scale, nesting, and distributed paradigm for information required suggests a specific watershed model framework be used as a synthetic template, analogous to the Everglades Landscape Model (ELM) and South Florida Water Management Model (SFWMM) which are used to evaluate management and

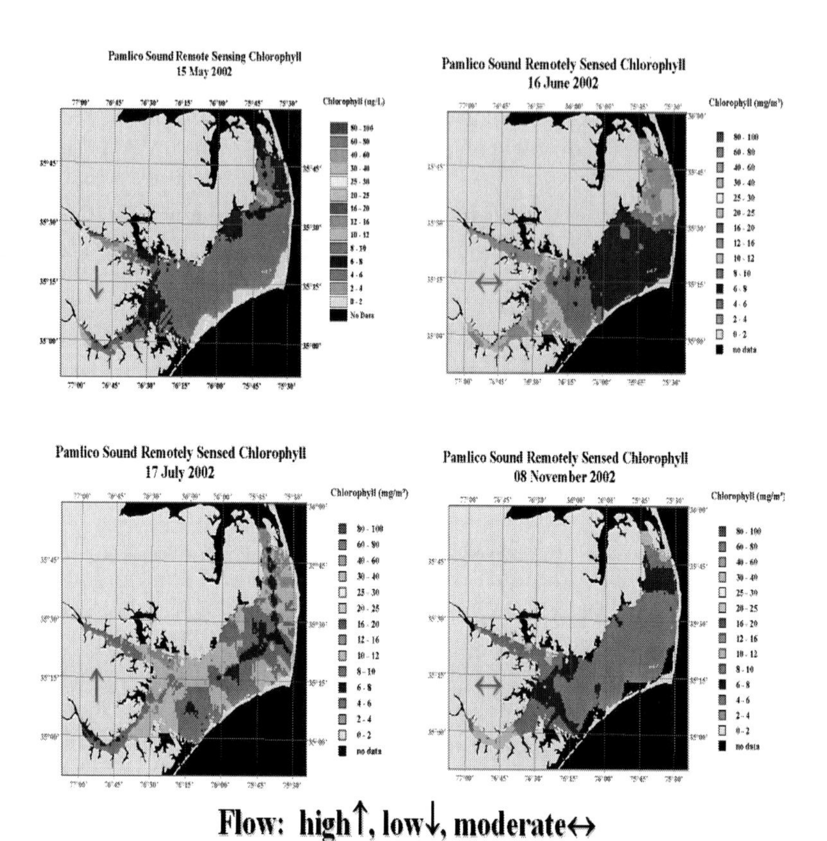

FIGURE 4-16 Spatial relationships between the phytoplankton biomass, as chlorophyll *a*, and freshwater discharge in the Pamlico-Albemarle Sound estuarine system. Surface-water chlorophyll *a* concentrations were estimated using an aircraft-based SeaWiFS remote sensing system (Courtesy L. Harding, University of Maryland Center for Excellence in Service), calibrated by FerryMon and ModMon samples. Under relatively low flow and long residence time conditions, phytoplankton biomass is concentrated in the upstream reaches of the estuaries. Under moderate flow, phytoplankton biomass maxima extend further downstream. Under high flow (i.e., short estuarine residence time) phytoplankton biomass maxima are shifted further downstream into the open Pamlico Sound. SOURCE: Reprinted, with permission, from L. Harding, University of Maryland.

TABLE 4-1 Nested Sensor Design With In-Situ and Remote Sensor Systems Covering Six Orders of Magnitude of Hydrologic and Biogeochemical State and Flux Variables

Length scale	10^{-1}-10^{1} m point – plot	10^{2}–10^{3} m Catchment	10^{4}–10^{5} m Regional watershed
Soil moisture and tension	time domain reflectometry (TDR), tensiometer Depth nested through rooting zone	TDR, tensiometer wireless sensor networks	microwave, thermal remote sensing
Soil solution chemistry	Tension lysimeters, nitrate microsensors[a] through rooting zone drainage lysimeters below rooting zone	Nitrate microsensors[a], redox probes coupled to wireless sensor networks Hillslope transects, oversampled in riparian zones	
Groundwater level, chemistry	Transducer, grab samples, nitrate microsensors[a/] Wells screened to multiple depths	Well clusters in wireless sensor network—transducer network Hillslope and flowpath transects	GRACE
Evaporation transpiration	Evaporation pan, Bowen ratio station Standard station height, Below canopy and in clearings	Eddy covariance Tower flux systems in multiple land-covers	MODIS visible/NIR and thermal remote sensing estimation evaporative fraction GOES estimation of net radiation
Precipitation and precipitation chemistry	Precipitation gage throughfall collectors wet/dry collector Paired gages	Precipitation gage clusters throughfall collectors wet/dry collectors	Polarimetric radar, precipitation gage and disdrometer network
Runoff and streamflow quantity and quality	Runoff plots lab analysis of storm event water chemistry	Runoff plots in multiple land use, soils, topographic settings Nitrogen isotopes	Stream gage network sampling channel flow/chemical/sediment concentrations from catchments ranging from 10^{-1}–10^{4} km^{2}, stratified by major physiographic province/land use, altimetry Real-time reagent analysis, turbidimeter, nitrate microsensors[a] Monthly lab analysis of fuller chemical and sediment concentration

TABLE 4-1 (continued)

Microbial ecology	Hyporheic zone and soils: cell counts with DAPI, enumeration of target taxa with FISH[b/]	Downstream transects: cell counts with DAPI and enumeration of target taxa with FISH[b/]
15N analysis	15N addition experiments to small hyporheic zone subplots [c/]	Natural 15N and 18O analysis of stream samples[d/]

NOTE: NIR = Near Infrared; GOES = Geostationary Operational Environmental Satellites.

[a]As per Bendikov et al. (2005); see Chapter 2. Additional micro/nanosensors could be incorporated as they become available and are field tested.

[b]A suite of different microbial ecologic methods may be used, most likely to include the DAPI staining method of Porter and Feig (1980) to determine total cell counts along with fluorescent in-situ hybridization (FISH) for phylogenetic identification without colonization (Amman et al., 1995).

[c]E.g., Mulholland et al. (2000), Hamilton et al. (2001).

[d]See Kendall (1998).

restoration strategies in South Florida. Given the watershed characteristics and the nature of the management problems, these models should have the capability to simulate hillslope and stream reach routing and biogeochemical transformations. This class of model systems would include (but are not limited to) RHESSys (Band et al., 1993, 2001; Tague and Band, 2004), the ecosystem land models (see Costanza and Voinov, 2004, of which the ELM is one example), and a set of distributed hydrologic models to which carbon and nutrient dynamics could be added such as distributed hydrology soil vegetation model (DHSVM) (Wigmosta et al., 1994), Real-Time Integrated Basin Simulator (tRIBS) (Ivanov et al., 2004a, b), TOPMODEL-based Land Atmosphere Transfer Scheme (TOPLATS) (Famiglietti and Wood, 1994), and others. The set of models could also be used to develop instrument deployment strategies based on expected value of additional measurements and testing of simulated behavior as specific hypotheses. In all cases, simulation modules addressing biogeochemical transformation within the channel network would require refinement and coupling to the nested watershed modules at appropriate scales.

Estimation of Hillslope/Riparian Connectivity, Transformations, and Transport

Within the drainage areas of the small, gaged catchments as well as representative areas draining directly into larger reaches, sampling is designed to assess the connectivity and development of organized moisture and shallow groundwater patterns. These properties are seen as a key factor controlling runoff production (e.g., Western et al., 1999) and transport of nutrients (Band et al.,

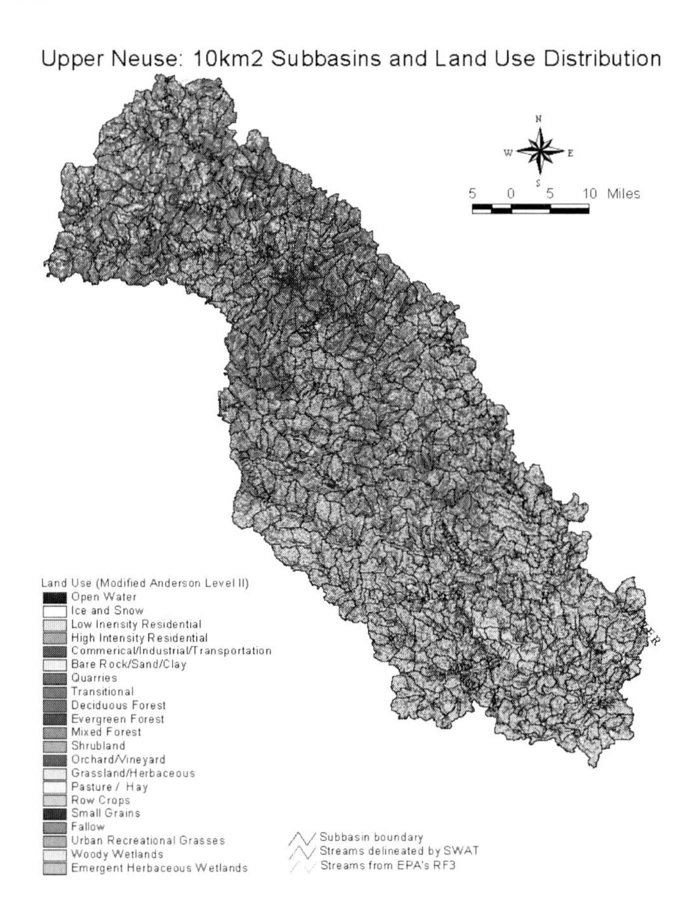

FIGURE 4-17 Land use for the Upper Neuse River Basin with 10 km^2 threshold subcatchment partitioning.

2001). Soil moisture and tension would be monitored at multiple depths along topographic flowpaths in representative land uses in each of the NRBE geomorphic zones. In order to assess the flushing mechanism hypotheses, sampling spatial and temporal frequency in wireless "intelligent" sensor network design would be designed to increase prior to and following recharge events as indicated by precipitation gage and radar networks. Sufficient sampling density would be carried out to estimate mean and variance of soil moisture conditions within areas commensurate to coarse resolution satellite sensors (e.g., MODIS TERRA and AQUA/AMSR) for calibration of remotely sensed moisture patterns.

Riparian groundwater wells at the downslope terminus of these transects would monitor groundwater level and gradients to determine connectivity and flux of hillslope or riparian water stores to the streams. Sensors recently developed to continuously monitor soil redox potential (e.g., Vorenhout et al., 2004) would be deployed in transects parallel to the wells, and both sets of sensors would be configured as part of the wireless soil water sensor system. These wireless and automated sensors would be complemented by periodic tracer experiments under different wetness and flow conditions designed to estimate hydrologic connectivity between upland, riparian, and stream/river channels of water and material constituents.

Additional geophysical tools need to be utilized to map soil and bedrock structure (e.g., ground penetrating radar) that influence the structure and transience of subsurface flowpaths. It is expected that these applications may be much more difficult due to current high uncertainty of subsurface flowpath patterns and high-frequency variations in actual patterns, but intelligent, wireless design may be useful to trigger the periodic mapping and sampling necessary to understand their transient dynamics. A particularly interesting application would involve the pore-scale nitrate sensor technology described in Chapter 2, extended to include other N species and coupled to the dissolved organic carbon and redox sensors to measure the timing, conditions, and rates of N transformations in upland and riparian nodes along transient hillslope flow networks.

Measurement Along Hydrologic Flowpaths

Novel assessment of groundwater/surface-water interaction along flowpaths in specific stream and hillslope flowpaths would extend the current set of sites to sample evolution of runoff and stream-water across the order-of-magnitude scale range. New technology for rapid, automated sampling of streamwater nutrient concentrations and transformations would be collocated with a larger subset of the stream gage network to compute concentration-discharge relationships and loads. These methods would be designed to detect and quantify denitrification and other transformation rates within reaches. Reaches would be chosen to sample dominant land-use and geomorphic conditions over a range of catchment drainage areas. Existing and emerging technologies that could be used include

1. Use of fiber optic cables along stream reach bottoms to estimate benthic water temperature with high precision and high spatial resolution (Selker et al., 2006) to map areas of upwelling tied to local hydrogeologic and fluvial geomorphic form and flow patterns to target measurements of water biogeochemical properties along the reach.
2. Instrumentation for real-time measurement of nutrient concentrations to monitor stores and fluxes of specific nitrogen species, as well as conservative trac-

ers through stream reaches as targeted by the stream bottom water temperature patterns and bed configuration. New technologies that have recently emerged include wet chemistry analyzers (e.g., http://www.lachatinstruments.com/products/qcfia), ultraviolet nitrate sensors (in-situ ultraviolet spectrometer—ISUS, MBARI), and multi-parameter sondes. In addition, field-deployed mass spectrometers are becoming available that would also determine oxygen and nitrogen isotopic signatures of evolving streamwater composition.

3. In larger, low-gradient river reaches, Membrane Inlet Mass Spectrometry (MIMS) can be used to gain integrated measurements of flowpath denitrification over heterogeneous reach conditions as demonstrated by McCutchan et al. (2003).

Characterization of the hyporheic environment in each of these reaches would be carried out to provide details of reach hydraulic and biogeochemical conditions associated with downstream evolution of streamwater chemistry and used to develop relationships with easily observed geomorphic and biological stream and riparian variables. Existing and emerging sensor technologies that would be useful for these purposes include

1. In-situ microcosms and continuous flow reactors to determine biogeochemical transformations of nitrogen and associated reactants.
2. Micro- and nanosensors to characterize biofilm composition and activity in hyporheic zones. These methods are still in development phase (especially for aqueous systems) but would provide the potential to estimate spatial and temporal variations in microbial dynamics in hyporheic and riparian zones within and between stream reaches.
3. Coupling of (1) and (2) above with characterization of microbial community structures in stream and hyporheic zone environments to determine the microbial effects on nitrogen cycling (e.g., Maurice et al., 2002).
4. Nitrogen enrichment experiments (e.g., Gooseff et al., 2004) and nitrogen isotope (N-15) tracer studies (e.g., Peterson et al., 2001) to determine rates and mechanisms of N cycling in streams and to differentiate autochthonous versus allochthonous N sources.

Summary

The design of a program following this case study integrates a nested measurement and monitoring system that samples at spatial and temporal scales appropriate to the coupling of WCN cycling, transport, and retention processes. In contrast to the Everglades case study, there is not a centralized institutional coordination for research and management in the Neuse, other than collaboration between federal and state agencies, as well as local universities and a set of nongovernmental organizations.

The geographic scale and complexity of this case study is intermediate to the typical principal investigator or small-team investigations and the national-scale environmental observatories (e.g., CUAHSI, CLEANER, NEON). It envisions a finite period of more intense instrumentation to test specific hypotheses regarding N sources, sinks, and transport, and to calibrate remote sensing and distributed modeling of WCN dynamics in the NRBE, followed by less-intensive but long-term in-situ instrumentation, remote sensing and modeling methods in support of watershed management.

To summarize, the NRBE is currently the site of a comprehensive set of measurement and monitoring efforts of federal, state, and local agencies, as well as the several universities in the area. To integrate and coordinate these efforts into a framework to better support the major hydrologic science and management issues, additional activity could include

- Development of an integrated information system that could access information generated by the diverse set of research and management entities and synthesize the information into a framework emphasizing spatiotemporal trends in WCN storage, flux, and residence time.
- Further development and integration of a nested monitoring and measurement system that incorporates in-situ sensor networks and is adaptable to spatial and temporal events.
- Parallel evolution of spatially distributed modeling approaches capable of resolving processes at scales commensurate with the sensor network and designed to characterize both continuous and episodic events. These models would serve as conceptual and operational frameworks for watershed system behavior, would be continuously tested and evolved by the observational network, and would be designed to assimilate information generated by the full sensor network estimation of system state and flux variables.
- Incorporation of human individual/institutional activity as integral components of the watershed. This is necessary to develop a management model capable of forecasting both short-term and long-term response of environmental policy and regulatory activities. The instrumentation technologies discussed here would need to be matched by social science research instruments to investigate controls and factors influencing individual and institutional behavior relative to WCN cycling and transport in multiple societal and geographical sectors.

CASE STUDY VI—MOUNTAIN HYDROLOGY IN THE WESTERN UNITED STATES

A number of recent documents, including *Water 2025: Preventing Crises and Conflict in the West* (U.S. Department of the Interior, 2005), highlight the need for new water information to enable better decisionmaking for water resources management, and for the myriad of other decisions that are influenced by water. Mountains are the primary source regions for water across the West. Challenges of complex topography, limited access, and large distances in mountains mean that new approaches to measurement of hydrologic properties, beyond those now in use, are needed to estimate spatial hydrologic variables and drive the next generation of predictive models. Integrated measurement systems that combine representative local but accurate ground-based data with broad spatial coverage from satellite remote sensing will be essential.

The demand for water information is large and growing. Western states, more than the rest of the country, depend on the judicious allocation of their water resources to support their industry and increasing population. For example, rivers draining to the California Bay-Delta, through a network of reservoirs and other hydraulic works, provide two-thirds of California's drinking water, support the $1.4 trillion state economy, and irrigate 7 million acres of the world's most productive farmland. The region is home to 130 species of fish, 225 species of birds, 52 types of mammals, and 400 plant species (California Bay-Delta Authority, 2000). In addition, the network of reservoirs on rivers draining to the Delta protects millions of people in major California cities against flood disasters.

Real and Emerging Problems

Explosive population growth and changing climate have combined to create mismatches between water supply and demand across the West. Mountain snow pack, the main source of the West's water resources (Figure 4-18), is particularly vulnerable. Water stresses are pervasive, affecting freshwater and coastal water quality, ecosystem/forest health in Arizona, Sacramento- San Joaquin Delta levee failure, groundwater mining for irrigation in California's Central Valley, and forest-fire occurrence across western forests, to highlight just a few phenomena. As water becomes a more valuable commodity, more accurate information than is currently available to support estimates of natural-water reservoirs (e.g., snow pack, groundwater), understanding of fluxes (e.g., evapotranspiration, groundwater recharge), hydrologic forecasting (e.g., water supply, floods, droughts), and decisionmaking is not just essential but critical. The technology in current use on these problems is decades old, and the blueprints for modernization are in most cases lacking.

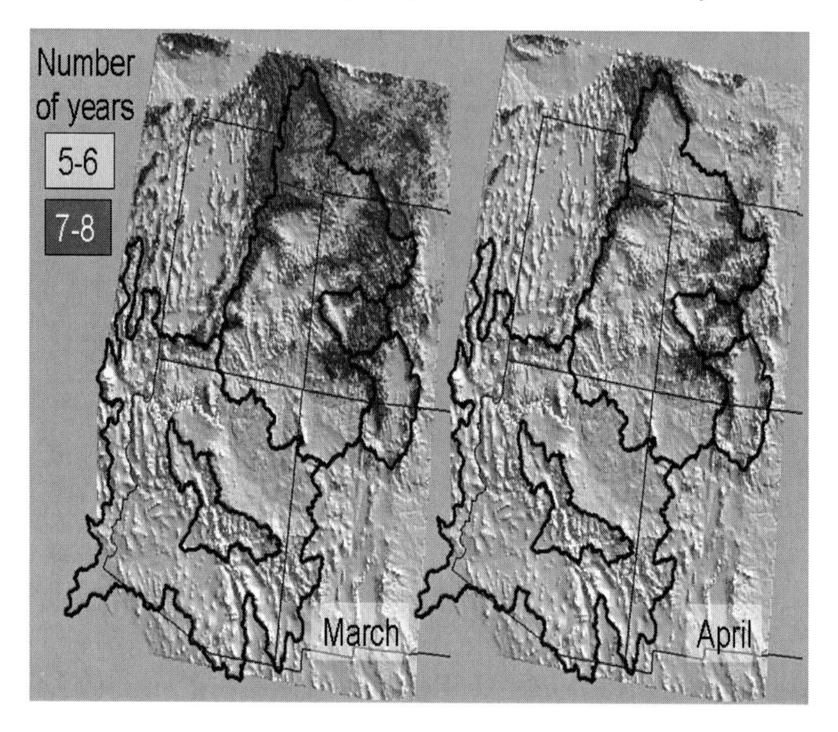

FIGURE 4-18 Persistence of snow-covered area in the Colorado River Basin, western United States over the 1995-2002 period for March and April, reported as number of years each 1 km^2 Advanced Very High Resolution Radiometer (AVHRR) pixel had detectable snow cover during that month. Snow pack, which covers a relatively small fraction of the basin, provides over 85 percent of the annual discharge in the Colorado River. SOURCE: Reprinted, with permission, from Bales et al. (2008). © 2008 by the American Geophysical Union.

While research challenges span water issues in the West, the potential for integrated observation systems involving new measurement technologies to impact these issues is perhaps greatest in mountain hydrology (Bales et al., 2006). In the mountains of the western United States, sharp wet-dry seasonal transitions, complex topographic and landscape patterns, steep gradients in temperature and precipitation with elevation, and high interannual variability make hydrologic processes and variations significantly different from lower-elevation regions or those that are humid all year. Hydrologic feedbacks in mountainous regions control the availability of water, influence the distribution of vegetation, dominate biogeochemical fluxes, and contribute to global and regional climate variability. Snow in mountains of the West is the main source of the region's water, with downstream hydrologic processes (e.g., groundwater recharge) and interactions with ecosystems controlled by processes at higher elevations.

Despite the importance of mountain regions to the hydrologic cycle, the processes controlling energy and water fluxes within and out of these systems are not well understood. Further, the lack of integrated measurement strategies and data/information systems for hydrologic data hamper improvements. As examples, we lack a robust framework for accurately describing and predicting the partitioning of snowmelt into runoff versus infiltration and into evapotranspiration versus recharge (Figure 4-19), and we lack strategies to exploit emerging technology to more accurately measure the spatial variability of snow cover and soil moisture in the mountains. The volume of mountain-block and mountain-front recharge to groundwater and how recharge patterns respond to climate variability are poorly known across the mountainous West (Earman et al., 2006; Wilson and Guan, 2004).

Three aspects of the mountain water cycle in the western United States are used to illustrate the needs and opportunities: (1) precipitation and microclimate, (2) snow pack, and (3) soil moisture. In each of these, researchers must determine the optimal sampling strategy based on such considerations as cost versus resources, reliability, effectiveness, variability of the parameters in question, sensitivity of models to the given parameter(s), potential for new methods to make a substantial impact, and needs of regulators and managers. One important application for this new knowledge is illustrated, hydrologic forecasting, including the water/energy cycle coupling and data integration needed for the emerging generation of forecast tools.

Microclimate and Precipitation

Microclimate can vary substantially at a scale of meters in the mountainous West, with large diurnal fluctuations. Direct measurements of precipitation in mountain environments are particularly challenging, because of the need to cover a large range of elevations and orographic positions. Mountains often have too much topographic variability to effectively use the Doppler Radar systems that have proven useful for monitoring extent and intensity of precipitation throughout the Midwest. Moreover, much of the precipitation falls as snow, and precipitation gages catch too little of the snowfall. Although it is possible to infer snowfall rates from snowpack accumulation measurements, direct observations of precipitation are needed for research (for example, to explore changes in precipitation type [e.g., Knowles et al., 2006] associated with climate variability and change), and for applications, such as to drive flood forecasts.

The spatial patterns of precipitation in mountainous terrain are nearly impossible to measure at the resolution of basin-scale hydrologic models (e.g., ~1 km). The primary ground-based resources available to estimate these patterns are several existing, overlapping networks, including NWS cooperative stations, RAWS (Remote Automated Weather Stations), USDA SNOTEL (Snowpack Telemetry) stations, and some smaller networks. Most of these stations lack the

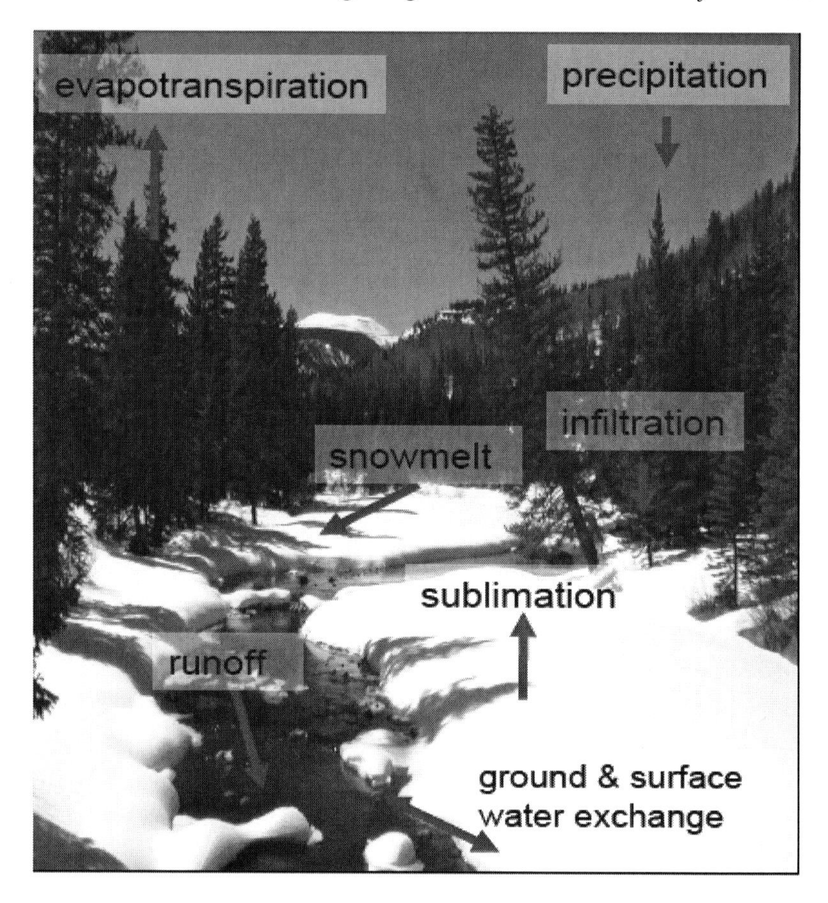

FIGURE 4-19 Schematic of the inter-related fluxes comprising the mountain water cycle, and partitioning of snowmelt. Photograph courtesy of Noah Molotch, University of California, Los Angeles.

capacity to differentiate snow from rain. The National Weather Service installed the Next Generation Radar system (NEXRAD) in 1994 to improve operational measurements of precipitation around the country. However, NEXRAD signals are occluded by mountains and thus are less reliable in the complex terrains where snowfall occurs. Even where precipitation is measured in situ, wind effects limit measurement accuracy. Traditional precipitation gages catch too little snow and cannot discriminate solid from liquid precipitation; when they catch snow, the measurement typically registers when the snow caught by the gage melts, not necessarily when it falls, causing a temporal lag.

Satellites offer an advantage over the operational radars in producing spatially distributed estimates in mountainous areas due to the unobstructed field of view. The high spatial and temporal resolution of precipitation estimates and the short latency of data availability make geostationary satellites the platform of choice for operational applications. However, microwave data from polar orbiting satellites or combinations of data from geostationary and polar orbiting platforms offer more reliable estimates for applications for which coarse resolution and long data latency are not of concern (Anagnostou, 2004). Distinguishing snowfall from rainfall remains a significant problem in satellite precipitation estimation, with current approaches relying on temperature thresholds or on the use of microwave satellite sensors (e.g., the Advanced Microwave Sounding Unit—AMSU).

A network for mountain precipitation, along with associated microclimate measurements important for determining energy fluxes and energy/water interactions, would ideally take advantage of both strategically placed ground sensors and satellite-based (or airplane) remote sensing. However, a major limitation is the availability of accurate precipitation sensors for remote deployments. Other components of microclimate should be measured spatially using a combination of a relatively few well-instrumented conventional measurement stations that offer low spatial resolution but multiple types of well-calibrated and conventionally accepted measurements, complemented by widely distributed embedded sensing devices that greatly increase spatial sampling resolution, but with cheaper probes and fewer types of measurements. For example, temperature and solar radiation (at a variety of wavelengths) probes can be attached cheaply to a microprocessor with radio transceiver and solar battery for wide distribution in a sensor network. For integrated, comprehensive, water balance measurements, at least some of the sensor nodes should be equipped with additional measurement devices such as wind speed and direction measurement; atmospheric moisture sensors, snow, soil moisture and temperature and other sensors. Installing these sensors in an embedded sensor network would help to control more complex sensing devices, to "flag" events in real-time for greater attention by researchers or managers, and to interact directly with models and control devices. Use of digital cameras to visualize the snowline and weather conditions, and potentially to read gages, is feasible provided icing can be controlled.

Scientists must often assess the trade-off between more expensive but potentially more labor-intensive measurements of a wide variety of parameters versus highly distributed inexpensive sensors for fewer parameters, some of which may be at lesser precision and accuracy. Given the physiographic variability of mountains, network design that encompasses both types of sampling strategies, along with remotely sensed information, is likely to result in significant success (Figure 4-20). It will be important to use remotely sensed data in the initial sampling network design, and as a potential independent method of data validation and verification. On the other hand, the sensing network should help to ground-truth remotely sensed information. Together with modeling, the combination of methods will provide a strategy for merging data at different scales, hence maximizing spatial and temporal resolution. Models can also help guide measurement design.

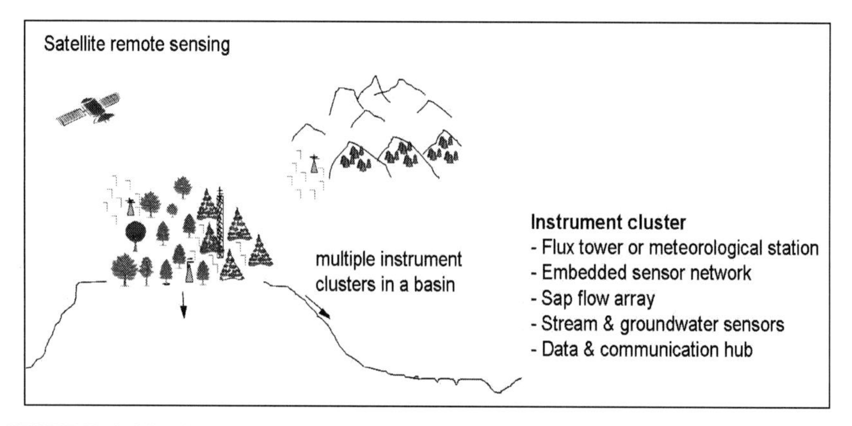

FIGURE 4-20 Conceptual design and deployment of instrument clusters in a mountain basin, integrating satellite remote sensing with strategically placed ground measurements. Selected instrument clusters are anchored by an eddy-correlation flux tower extending above the forest canopy, with ground measurements extending 1-2 km from the tower. Other clusters would consist of sensors and sensor networks but not a tall tower. Adapted, with permission, from Conklin et al. (2006). © 2006 by the American Geophysical Union.

Special consideration will need to be made for engineering embedded sensor pods for harsh and variable mountain conditions, and to designing the sensor network in concert with the topographic challenges (e.g., appropriate relay stations to get around obstacles to wireless communication). In some locations, there may be the need for sensor network data from an array of local sensors to be stored on-site for periodic download rather than telemetered directly to the observer.

Snowpack Properties

Hydrologic and land-surface models are increasingly including mass balance and physically based snowmelt models (e.g., Cline et al., 1998). While several of such models are being used by the research community, improvements in the treatment of snowmelt as well as other model components depends on data. For snowmelt, data needs include spatially distributed components of the surface energy balance, as well as snow properties (Figure 4-20). Ground observations of snow water equivalent (SWE) have been used in conjunction with remotely sensed snow-covered-area (SCA) data to estimate the spatial distribution of SWE across mountainous watersheds (Fassnacht et al., 2003). However, such efforts cannot be expected to yield representative measures of snow distribution across a basin owing to the non-representative location of SWE measurements (Molotch et al., 2006).

Several sources of seasonal snowcover data exist, ranging from information collected as part of weather monitoring to hydrologic data from networks dedicated to snow data collection, and more recently to remotely sensed products from polar orbiting and geostationary satellites. Remote sensing is the only practical way to measure the spatial extent and variability of snow cover and albedo, and, during the past decade, methods for mapping snow-covered area from visible and infrared instruments on satellites have become well developed (Dozier and Painter, 2004). However, none of the satellite data sets encompass enough system interactions to be considered "snow system" data (Bales et al., 2006). Snow water equivalent is measured at over 1700 points in the Western United States, from a combination of manual snow surveys and transmitting snow pillows. While this large number of samples provides regional knowledge of the spatial distribution of SWE, it is insufficient to resolve the variability of SWE and snowmelt at the basin scale (Figure 4-21). Moreover, most of snow courses and automated stations are situated on flat or nearly flat terrain, and are preferentially placed high enough to that they stay snow covered most of most winters (to justify the expenditures), which means that we are largely blind at altitudes where snow pack is more ephemeral. Also, there are no stations at the highest elevations, which contribute most of the late-season snowmelt. Glaciers, which also help sustain baseflow after seasonal snow has disappeared, are also severely undersampled. The existing snow measurements are used as *indices* of streamflow, rather as direct measurements of basin-scale snow volumes.

A comprehensive snow-measurement network would blend sparse but detailed, accurate measurements of snow water equivalent, microclimate, and other water-balance variables with satellite remote sensing and spatially extensive lower-cost measurements of snow depth and other low-cost sensors. The ground-based system would ideally be composed of low-cost sensors in embedded sensor networks (Figure 4-22) that build outward from existing ground-based snow pillows and snow courses to capture the physiographic variability across a catchment (Molotch and Bales, 2005, 2006). Building outward from existing measurement sites takes advantage of their long and valuable record of measurement, and provides the additional measurements needed to use distributed depth measurements. As snow density varies much less spatially than does snow depth (Molotch et al., 2005), many more depth than SWE measurements are needed. Low-cost temperature sensors can also be used to infer snow depth, and other technologies are under development. Satellite data can reliably and accurately provide SCA information. Various interpolation and modeling strategies are available to blend the ground-based SWE and satellite SCA to provide basin-scale estimates of SWE for predictive modeling. Periodic SWE measurements from aircraft platforms can also contribute to this mix (e.g., gamma ray sensors). Thus snow properties are an excellent example of how relatively modest investments in integrated observation systems can immediately provide critical data products.

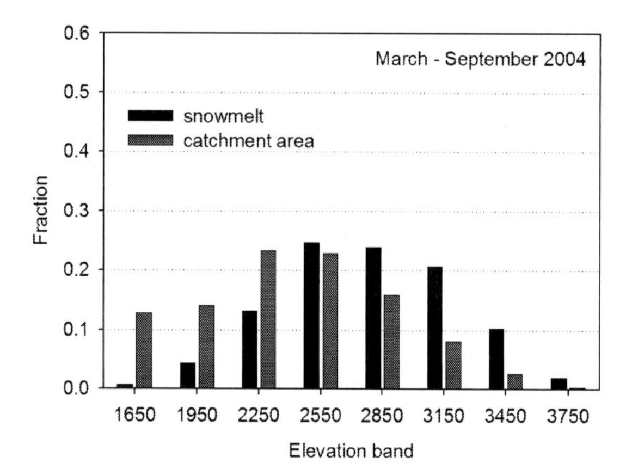

FIGURE 4-21 Contributions of various 300 m elevation bands to snowmelt in Merced River Basin, Sierra Nevada, California. Data derived snowcover depletion based on MODIS satellite data. Fraction of basin in each elevation band is given for reference. Continuous, ground-based snow measurements are limited to three sites in the basin (2100-2500 m elevation). SOURCE: Reprinted, with permission, from Bales and Rice (2006). © 2006 by the American Geophysical Union.

FIGURE 4-22 One pod in a sensor web that was installed to measure snow depth over a 20,000 m^2 study area in Yosemite National Park. Depth sensor is on right and wireless pod on left. Photograph courtesy of Margot Wholey Photography.

As satellite radar retrievals of SWE advance, they will and can be integrated into the spatial interpolation schemes developed around the in-situ network. However, for the foreseeable future, there are no operational radar satellites with sufficient spectral and polarization capabilities to infer snow water equivalent in the mountains. A ground-based observational network remains an important component of the snow observing system that will not be replaced by the satellite system because of inherent uncertainties in retrievals. Satellite-based snow cover/depth observations might cover the future needs of some users of snow data, but economics and scientific objectives now require a merging of all available snow information in an enhanced data set. Judicious and strategic extensification of in-situ measurements to lower and higher elevation sites coupled with advancements in remote sensing acquisitions will provide the means to a long-term, high-resolution monitoring of snow pack properties.

Soil Moisture

Soil moisture is a primary state variable of the land surface. In mountains, soil moisture is greatest following spring snowmelt and lowest in late summer and fall after the soil dries. Its spatiotemporal variability affects surface and subsurface runoff, modulates evaporation and transpiration, determines the extent of groundwater recharge, and initiates or sustains land surface-atmosphere feedbacks. Soil moisture is influenced by: (1) precipitation history, (2) soil texture, which determines water-holding capacity, (3) land-surface slope, which affects runoff and infiltration, and (4) vegetation, land cover, and bedrock slope/depth, which influence evapotranspiration and deep percolation. The partitioning of soil moisture to groundwater recharge, evapotranspiration, and surface/subsurface runoff at different spatiotemporal scales and under different hydroclimatic conditions poses one of the dominant challenges in quantifying water cycle variability (Jacobs et al., 2006).

Soil moisture is measured at some RAWS and SNOTEL sites, but these generally do not lie in mountain settings. Remote sensing of soil moisture is made in the microwave frequencies.

As in the case of SWE, passive microwave retrievals of soil moisture are too coarse for the spatial variability and rugged terrain in mountain regions. Radar retrievals of soil moisture (and SWE) are in research mode now but again the lack of an operational radar satellite presents a significant obstacle to the implementation of these retrievals (Shi et al., 2002; Oldak et al., 2003; Western et al., 2004). Moreover, rugged terrain and vegetation cover can confound retrievals. While the technical difficulties are great, definition of the spatial variability of soil moisture is critical in modeling hydrologic response in mountain catchments (Zehe and Blöschl, 2004).

The strategy for measuring soil moisture in mountain catchments mirrors that for snow depth, that is, design a network that captures the spatial variability in physiographic features (e.g., north- versus south-facing), plus distance from trees. A number of low- to moderate-cost soil moisture sensors are available that can be incorporated into sensor networks, and new technology is emerging. As radar retrievals of soil moisture advance, they should be integrated into the spatial interpolation schemes developed around the in-situ network. Like SWE, proper network design for soil moisture measurement remains a research issue.

Data Integration and Applications

An additional, more-general challenge that cuts across all aspects of water in the West is that of data integration. Current computing environments, investigator-specific research practices, and agency data distributions are disjoint and do not readily facilitate system/data integration. Typically, these systems use ad hoc scripts to perform the required processing and idiosyncratic naming conventions for the files that hold the products. Data extraction from a variety of systems is, therefore, time-consuming and subject to error proliferation, especially when assembling a synoptic view or parsing the data according to a suite of criteria (e.g., data from all snow courses above some particular elevation in a particular basin with more than some threshold length of record). Our current modes of analysis usually require reorganization of data and creation or rediscovery of metadata values for each product. Dissemination, especially where custom processing such as subsetting, reprojection, or reformatting is required, is often treated in a similarly ad hoc fashion. The technologies to solve most of these problems are at hand already, but implementation will require a concerted, collective commitment by users and providers (Bales et al., 2006).

Cyberinfrastructure advances can overcome many of the current data problems by making data and information available in ways that are convenient for users. That does not necessarily mean that data are made available to users in the same way that they access data now. Rather, technologic advances that are tailored to be responsive to community needs can both make users more cyber-savvy and information more accessible.

Hydrologic forecasting is used to illustrate the role of new, integrated measurement systems. Hydrologic forecasting methods for operational management are well established, and are based on several decades of historical data from sparse networks that monitor surface precipitation and temperature, snow pack, and river stage or discharge. Specifically, flow-forecast reliability and accuracy depend critically upon the use of historical data to calibrate the operational hydrologic models for the watersheds of interest (Fread et al., 1995; Finnerty et al., 1997) and upon the use of reliable quantitative precipitation forecasts (Olson et al., 1995; Sokol, 2003). However, forecast skill for spatially distributed flow forecasts over small areas in the populous coastal mountainous basins or for seasonal forecasts of reservoir inflows in large reservoir projects on the Sierra Nevada is in many cases poor for effective emergency and water supply management.

Long-range reservoir inflow forecasts depend largely on long-range forecasts of surface climatic variables (temperature and precipitation). Nevertheless, reliable estimation of forecast uncertainty for such long-range forecasts will benefit directly from improved good quality hydrologic observations, and recent integrated forecast-management demonstration projects show that reliable forecast uncertainty is necessary for improved reservoir management (Yao and Georgakakos, 2001; USACE, 2002; Georgakakos et al., 2005). The improvement of flow forecasts over small areas will require improvements in high-resolution measurement technology for precipitation and temperature (NRC, 2005) and the use of real-time flow-forecast assimilation procedures (Seo et al., 2003). Here again, good quality data will contribute directly to improved emergency management effectiveness. In addition, future conditions (e.g., warmer temperatures, snow-to-rain transitions) are expected to be outside the range of past system behavior. Several recent studies find evidence that climatic impacts on western water resources are in transition to a new regime (Stewart et al., 2005). This would make large improvements in hydrologic observations and information even more important.

Summary

Mountain snow pack is the main source of the American West's water resource. However, this resource is particularly vulnerable to ongoing spatial and temporal changes in melt patterns, which will directly affect the seasonal availability of water to the multitude of stakeholders in the region. Three aspects of the mountain water cycle in the western United States were used to illustrate the region's needs and opportunities: (1) precipitation and microclimate, (2) snowpack, and (3) soil moisture.

The integration of many types of observations can assist the management of mountainous water resources. In most cases, the resulting data and information need to be continuous, reliable, and rapidly available for effective prediction and management. For these reasons, effective strategies such as the use of embedded sensor network technology would contribute greatly to the progressive use of new data sources and types in the water resources management process and for quantifying their benefits and associated costs. The organization and careful monitoring of prototype demonstration projects with the participation of measurement specialists, scientists (including communication and computer scientists), forecasters, and managers is an effective means to develop and test such strategies.

To summarize the chapter, the six case studies presented above are located in climatic regions from subtropical to semiarid to alpine highland to circum polar. They range from large scale to medium scale, and from programs that are largely ongoing to initiatives that are nascent or proposed. Some are more oriented toward scientific understanding; others are oriented toward improving management decisions. Because of this, they provide a wide variety of lessons for government agencies, academic institutions, and even the private sector. These lessons are summarized and discussed in the following chapter.

5

Synthesis, Challenges, and Recommendations

This report offers a broad review and vision of integrated observing for the hydrologic and related sciences. As was articulated in Chapter 1 and the case studies in Chapter 4, population growth and global climate change will increasingly strain the use of fresh water for human activities and require greater efficiencies in our utilization of this essential resource. The difficulties in quantifying many of the stores and pathways of water, and the related energy and biogeochemical fluxes through the environment add layers of uncertainty to the problem. Since these fluxes vary over time and space scales, the integration of observations, data management, and process-based predictive models is essential to provide the information needed to expand our understanding of the linkages among the water, energy and biogeochemical cycles in both pristine and highly modified areas, and to provide useful information for water and environmental managers.

THE VISION

The vision is of the future where on-site ("in-situ") sensors that measure properties such as temperature, soil moisture, and water quality at high spatial and temporal resolution are integrated to form an embedded network that is in turn connected to other networks of observations. These other networks may consist of traditional observation platforms such as ground-based precipitation gages, river discharge gages, or "grab samples" for water quality, or remotely sensed measurements from airborne or spaceborne sensors at larger spatial scales. The vision includes the ability to integrate these measurements across all relevant scales and with models to offer wide-ranging predictive capabilities, and to make both data and predictions widely available through "web portals"—that is, through Internet-based access where users of the predictions and measurements such as water managers and educators could easily access the information.

Measuring physical, chemical, and biological properties at multiple scales will increase our process-based and predictive understanding and improve management. An excellent example of this is provided in Chapter 4 by the Neuse River Basin Study, because nitrogen cycling in that basin is controlled to a large extent by microscale processes in regions such as the hyporheic zone, yet understanding actual ecosystem dynamics and effects on estuarine systems requires the broad view provided by spaceborne remote sensing.

A PROMISING BEGINNING

While this vision appears futuristic, the study found—and the report documents—that *many elements of this vision currently exist.* For example

- *Sensors:* Chapter 2 presented a comprehensive discussion of sensors at a wide variety of scales that are currently under development and being tested in experimental observatories or in research projects that focus on particular environmental variables. This chapter also identified approaches that are emerging in other disciplines that may lead to additional new environmental sensor development, and that will need to be modified for actual field deployment to address issues in water and environment. These advances in measurement technologies range in scale from nanosensors focusing on biological variables to airborne sensors that offer spatial context for point measurements to satellite sensors that offer regional-to-continental scale perspectives.
- *Embedded sensor networks:* Embedded sensor networks, consisting of spatially distributed sensor-containing platforms connected to and often controlled by computers, and with the sensors themselves often containing microprocessors, are also being demonstrated. For example, as presented in Chapter 2, wireless but interconnected instrument buoys at remote lakes in Wisconsin have provided frequent measurements of lake quality. Dense measurements of hydrologic and meteorological sensors at the Santa Margarita Ecological Reserve are providing real-time data to drive watershed models. And sensors embedded in sewer systems are helping to combat combined sewage overflow.
- *Other communications and cyberinfrastructure:* In Chapter 3, the NSF-supported CUAHSI project development of "web portal" applications was presented along with one current example using this technology, the Central America Flash Flood Guidance (CAFFG) System. The importance of cyberinfrastructure was highlighted in several of the case studies presented in Chapter 4.
- *Modeling and data assimilation:* Chapter 3 also reviewed the considerable advances in data assimilation, that is, the merging of data across scales from in-situ point measurements to coarser scale airborne or satellite observa-

tions for use in models. Data assimilation has long been used in applications such as operational weather prediction and in ocean modeling, and its application in land hydrology and environmental science is being demonstrated in research applications.

- *Existing experimental watersheds, planned observatories, and international initiatives:* The USDA and others have many monitored catchments that have been and can continue to be used for basic science related to watershed processes. While these are generally at a scale too small to integrate satellite observations, they can be incorporated into the environmental observatories planned by the National Science Foundation. At the international level, the intergovernmental Global Earth Observation System of Systems (GEOSS) has been conceived and has strong international support.

Despite the great promise of integrated hydrologic measurement, many challenges stand in the way of its full development and application. Many of these challenges are illustrated by the case studies offered in Chapter 4. These studies cover a diverse range of current hydrologic research and management efforts, encompass a variety of related sciences, and provide detailed discussion on the need for integrating a wide variety of measurements with models for improved process understanding and management. While these case studies are not meant to be exhaustive in discussing integrated observations, they are representative of the breadth of opportunities facing the community. The most striking theme from the case studies is that *the gaps between the vision of what the researchers or managers want to achieve and their ability to realize that vision are real, but in many cases extremely narrow.* However, there are major challenges, which are listed below, followed by recommendations for overcoming these and other challenges.

MAJOR CHALLENGES

Despite the promising advances enumerated above, there are at least nine major challenges that need to be overcome before the vision for integrated observations described in this report can be brought closer to reality. These relate to

- Developing appropriately scaled sites for water science;
- Developing and field deploying land-based chemical and biological sensors;
- Inspiring a greater agency commitment to developing airborne sensors;
- Developing both new spaceborne sensors and a "research-to-operations" strategy for existing ones;
- Bridging the gap between sensor demonstration and integrated field demonstration;

- Integrating data and models for operational use;
- Adopting new integrated hydrologic measurement and modeling systems for water resources applications;
- Funding interdisciplinary science; and
- Addressing the fractured federal responsibility for hydrologic measurement, monitoring and modeling.

These challenges are described in more detail in the following paragraphs.

Developing Appropriately Scaled Sites for Water Science

Most existing field-based watershed research programs focus on small catchments—from <1 to (rarely) several hundred square kilometers—and most of the larger ones are not extensively monitored. They are managed by the U.S. Department of Agriculture (USDA) Agricultural Research Service and Forest Service, the U.S. Geological Survey (Water, Energy, and Biogeochemical Budgets, or WEBB, program), the Environmental Protection Agency (such as the Chesapeake Bay Program), and the National Science Foundation (Long-Term Ecological Research, or LTER, program). A large quantity of historical data exists from these sites, but much of it is difficult to access; even within the USDA these data are not in a single database.

Critically, few of these field observatories envisioned a major role for remote sensing when they were conceived, in some cases because of their small footprint, in others because they were designed before modern remote sensing methods were developed, and in still others perhaps simply because of institutional barriers. They generally succeed well in supporting the small-scale catchment science that they are designed for, but need to be about two orders of magnitude larger to approach scales that are appropriate for land-atmosphere coupling.

The proposed hydrologic, ecological, and water-quality "observatories" such as those of the National Ecological Observatory Network (NEON) and the WATer and Environmental Research Systems Network (WATERS) are designed to be much larger—on the scale of 10^4 to 10^5 km. Direct estimates of varying quality are now available for variables such as snow extent, snow water equivalent, surface-water height, soil moisture, total water storage, and precipitation; indirect estimates for groundwater, evapotranspiration, and streamflow can also be made in some cases (see Chapter 2, Spaceborne Sensors section). Not all of these estimates (e.g., groundwater) can be made at the scale of the proposed observatories, but improved sensors are proposed for many of these variables for the future (NRC, 2007), and data assimilation and modeling techniques are advancing as well.

There are, therefore, both unprecedented opportunities and unprecedented challenges for incorporation of space-based and airborne observations in observatories—and not as an afterthought, but as an essential part of the experimental design.

Development and Field Deployment of Land-Based Chemical and Biological Sensors

As discussed in Chapter 2, physical sensors, such as those that measure air and water temperature and pressure, radiation, and wind speed and direction, have evolved over decades and are now mass produced and routinely packaged together in small instruments along with power and communication devices. However, sensor development for many important chemical and biological measurements is relatively immature. Chemical sensors are needed to measure a wide range of elements and inorganic and organic molecules, in all environmental media. Biological sensors can provide key information on the function and structure of biologically influenced ecosystems in real time (NSF, 2005). For widespread use in the field, chemical and biological sensors would need to be inexpensive; robust against environmental stresses, such as temperature extremes and biofouling; have stable calibrations or be capable of remote calibration; and have low energy requirements. Development of a wide range of field-robust chemical and biological sensors is one of the greatest challenges facing widespread deployment of sensor networks in the hydrologic sciences.

Airborne Sensors

As discussed in Chapter 2, airborne and spaceborne sensors are needed in integrated observing systems to extend observations beyond the point measurement scale, with airborne measurements at a spatial scale that fills the gap between the in-situ plot-scale observations and the larger satellite-scale observations. As discussed in Chapter 2 and in National Research Council (2007), airborne remote sensing at the National Aeronautics and Space Administration (NASA) historically has been viewed as an intermediate step between initial sensor development and space deployment. Measurements from such airborne systems were collected during limited-duration field campaigns to help develop retrieval algorithms. Airborne sensors are also used to under-fly new satellite sensors for their initial validation. In these ways, NASA views its airborne science program as supporting its satellite sensor development program and not as a sensor program in its own right. This approach has unfortunately impeded the development of operational airborne observing platforms that could play a very important role in hydrologic observations.

From the above discussion, as illustrated by the case studies in Chapter 4, the critical challenge to address is the limited commitment of NASA and other federal agencies to airborne platforms as effective operational measurement systems, leading to a paucity of programs to develop smaller, less-expensive sensors that could be used on these platforms.

Spaceborne Sensors

In the area of satellite-based remote sensing, NASA has made good progress in developing and deploying a wide range of sensors for hydrologic science that are used primarily for research. Most of these are described in Chapter 3. Addressing the future needs in this area was the charge of another NRC committee (the "decadal survey"; NRC, 2007), whose relevant results are summarized in Appendix C. The missions that study recommended—including missions to measure diurnal precipitation, soil moisture, water storage in lakes and wetlands, and snowpack water storage, especially in mountainous regions—are consistent with the vision, findings, and recommendations of this study.

Nonetheless, two related challenges are relevant to this report: (1) for NASA, how to transition research sensors (and their costs) to operational agencies, and (2) for other (non-research) users, determining how to utilize these observations for operational uses. In the case of weather-related sensing, Congress has provided guidance. For other agencies and users, this remains a challenge. As an example, the loss of high-resolution (<100m) thermal imaging on the Landsat TM satellite and the decision by NASA not to have such a capability on the replacement satellite has affected agencies such as the USDA and state agencies concerned with irrigation water management that use this information operationally.

Thus, the critical challenges to address here are (1) a resolution of the "research-to-operations" transition from NASA-developed "experimental" satellite observations to the broad variety of operational agencies and users that need routine (i.e., operational) observations, and (2) the lack of a corresponding monitoring strategy by federal and state agencies (for example, the Environmental Protection Agency [EPA], USDA, the National Oceanic and Atmospheric Administration [NOAA], and state water and natural resources agencies) that would incorporate airborne and/or satellite remote sensing measurements, where appropriate (various case studies in Chapter 4 offer examples). These apply to existing but underutilized data sets as well as to future data sets.

Bridging the Gap between Sensor Demonstration and Integrated Field Demonstration

The last section of Chapter 2 (see Figure 2-10) presented the steps needed to advance sensors from experimental development through operational deployment. Also shown in the figure are various agencies that are currently involved in the different steps of the process or in complementary activities. While Figure 2-10 shows an idealized summary with overlapping agency activities, in reality there are significant interagency gaps between the steps of sensor development, sensor demonstration, integrated field demonstrations, and operational deployment of sensors. The greatest gap is between sensor demonstration and integrated field demonstration. Closing this gap would involve integrating the sensor networks and webs within hydrologic observatories and experimental demonstration sites, and interfacing the sensor networks with the broader development of cyberinfrastructure.

Integrating Data and Models for Operational Use

The importance of data-model integration is apparent in a number of the case studies. For the Mountain Hydrology study in Chapter 4, predictions of water availability are made from point measurements and model forecasts, leading to management decisions. In the Neuse River Basin Study, the management decisions are based on sparse water-quality measurements, largely limited to the main stem and high order tributaries. In these two case studies, both models and observations are used to guide management decisions; in each case a data assimilation system that merges models and observations would offer improved predictions for decisions. However, the systems required would likely be very different, as they would be for many other critical water resources problems. The challenge is to develop data-model integration methods that will be useful for broad families of applications, rather than just a few of the many possible applications.

The Next Step: Water Resources Applications

In the United States, large water resources problems involve multiple stakeholders, including government agencies, business interests, and the public. Management is typically diffuse, although the benefits of coordinated management are well recognized. Standard measurement and modeling techniques and rules for water management are entrenched and often legally mandated. For example, as mentioned in the case study on Mountain Hydrology in the western United States, the principal methods for monitoring snow pack and predicting snowmelt volumes have changed little over the past 100 years. This has the benefits of producing a consistent data set to show trends over time, and of simplifying training and daily tasks of staff. However, it also leads to missed opportunities to improve the accuracy and precision of the data and resulting model predictions. Changing modes of measurement and management will be extremely difficult, even as the nation's water resources challenges increase over time.

Funding Highly Interdisciplinary Science

Interdisciplinary science is much more common than it was 20 years ago, and it is no longer uncommon to see scientists and engineers from a variety of fields working on related problems in experimental watersheds. However, the design and use of integrated hydrologic measurement systems in specific research applications adds an extra layer of complexity to the challenge. Thus, these new kinds of projects will require unprecedented interdisciplinary cooperation, including interactions among scientists, engineers, field researchers, modelers, and theorists. Each application will be somewhat unique, requiring the participation of the technology developers (electrical engineers, computer scientists, and modelers) and the physical, chemical, and biological scientists who apply the technology to hydrologic research. While many universities and research laboratories have the required diversity of expertise, marshalling this expertise on specific projects will likely require new programs or sources of funding.

Addressing the Fractured Federal Responsibility for Hydrologic Measurement, Monitoring, and Modeling

The overarching barrier to the development and implementation of integrated hydrologic measurement systems is the lack of a single federal agency with primary responsibility for measuring, monitoring, and modeling the environmental factors and processes that control the hydrologic cycle. USGS is the agency with a mandate that comes closest to meeting this responsibility. It continuously monitors streamflows at over 7300 sites, as well as lake and groundwater levels at a smaller number of sites. It also collects data on stream water quality at about 2800 sites. Through its National Water-Quality Assessment (NAWQA) Program, USGS performs integrated assessments of a small number of watersheds. NOAA has the responsibility for monitoring precipitation and other meteorological variables and for monitoring and managing coastal waters. EPA has regularity authority for water quality and oversees the setting of water-quality standards on impaired stream reaches through the Total Maximum Daily Load (TMDL) Program. EPA also manages the Environmental Monitoring and Assessment Program (EMAP), which assesses national ecosystem health. Hydrologic modeling has been advanced by EPA, U.S. Army Corps of Engineers, the USGS, and NOAA—agencies with a wide variety of missions. Finally, NASA develops, deploys, and maintains satellite sensors that are used by others to assess the environmental factors, although its primary mission is space exploration.

Thus, it is easy to understand why the responsibility for measuring and monitoring the environmental factors and processes that control the hydrologic cycle is so fractured, given the evolution in our management of water resources and of our understanding of the hydrologic cycle. But the dual threats of global climate change and population growth demand a focused strategy for providing information on the nation's water resources and the environment.

RECOMMENDATIONS

Interagency Sensor Development

As shown in this study, there is significant development in new innovative sensors, especially in the area of chemical and biological sensors. Sensor development is taking place for a number of hydrologic processes and these new sensors are being tested within embedded networks. But these developments are not well coordinated, especially in their testing through field deployment. Linking sensor development to integrated field demonstration and operational deployment by state and federal water agencies needs to be fostered. Involvement of the private sector can help to ensure that sensor innovations are carried through the development phase and become incorporated in products that are widely distributed and economically viable.

There is an important opportunity for partnerships between the National Science Foundation (NSF), NASA, other agencies, universities, and the private sector in the development of new sensor systems. Such partnerships would enable the development of sensors and sensor systems that address multidisciplinary observational needs, including the needs of operational agencies. Interagency laboratories have been created in the past and have achieved some success, as in the case of the national sedimentation laboratory in Vicksburg, Mississippi.

Recommendation 1-1: *NSF, in partnership with NASA, NOAA, EPA, USGS, and possibly national health and security agencies, and with collaboration from the private sector, should develop one or more programs that address the need for multidisciplinary sensor development. An interagency sensor laboratory should be considered.*

Concepts such as the Consortium of Universities for Advancement of Hydrologic Science's proposed "Hydrologic Measurement Facility" (HMF) Center, which would house community instruments based on mature technologies, and HMF "nodes," which would be university-based, have a 3-6 year life cycle, and support a specific emerging technology, would complement but not replace such programs. In particular, universities are the ideal environment for training future users of environmental sensor networks and the data that will stem from them.

As described above for sensor development, and earlier in the challenges facing the development and deployment of integrated hydrologic measurement systems, the pace of advancement is constrained by the fractured federal responsibility for the development and deployment of integrated measuring, monitoring, and modeling systems, which seems to impede their utilization.

Recommendation 1-2: *Serious consideration should be given to empowering an existing federal agency with the responsibility for integrated measurement, monitoring, and modeling of the hydrological, biogeochemical, and other ecosystem-related conditions and processes affecting our Nation's water resources.*

This agency would manage the design, development, and deployment of integrated hydrologic measurement systems. Candidate agencies may include NASA, NOAA, and USGS because they all have extensive sensor and observations systems and modeling, and have the potential to take such a leadership role. It is recognized that taking on such a leadership activity would require new responsibilities, probably new organizations within the responsible agency, and new funding to carry out the mandate. There are some areas, such as cyberinfrastructure and the development of web-portal user interfaces, where expertise needs to be developed as suggested in other recommendations. There are many forms that such a lead agency activity could take, ranging from interagency oversight that utilized the services of existing agencies and private entities, to taking on operational responsibility for monitoring. The committee did not explore these details in any depth.

Observatories, Demonstration Projects, Test Beds, and Field Campaigns

Well-designed observatories, demonstration projects, test beds, and field campaigns have been and can continue to be effective for testing integrated observational-modeling systems to address research questions and operational needs. Many of the case studies presented in Chapter 4 clearly articulated the need for hydrologic observatories to demonstrate how the advances in embedded sensors can be integrated with traditional sensors, models, and remote sensing to provide new hydrologic process understanding. In fact, the successes achieved by the current experimental watersheds argue for more such projects.

Recommendation 2-1: *Coordinated and jointly funded opportunities for observatories, demonstration projects, test beds, and field campaigns should be significantly increased.*

The observatories provide mechanisms to improve knowledge of complex environmental systems and processes through improved sensor information and modeling. It is hoped that the observatories would have academic and government research scientists working together on problems that lead to enhanced management approaches by the agencies.

Generally, the funding of observatories, test beds, and field campaigns at the "integrated field demonstration" scale is challenging, because of the relative lack of funding at a scale greater than the individual research project and smaller than programs like NSF's Major Research Equipment and Facilities Construction (MREFC) projects. This is true even considering that costs can often be kept reasonable by encompassing projects within existing networks when appropriate to the research question. Likewise, most grant horizons and federal funding cycles (1-5 years) are inappropriately short for interdisciplinary, integrative projects.

Recommendation 2-2: *Agencies should consider offering new funding streams for projects at the scale of several million dollars per year for approximately 5-10 years to help close the gap between sensor demonstration and integrated field demonstration.*

These projects could be "problem-centric" with a significant research question or management goal. An integrated hydrologic observations network, as envisioned in this report, would be designed and deployed as an integral part of the project, and used to address the project goals and objectives. This class of project would help fill the gap between sensor development and operational use of established sensors. It could also be a critical demonstration mechanism to provide feedback to the more fundamental sensor development activities and offer results to guide operational applications of integrated sensors and cyberinfrastructure.

The problem posed in the Chapter 4 case study on water and malaria in sub-Saharan Africa illustrates this class of project. Its research question requires integration of data from a variety of sensor systems measuring quite different types of variables (hydrologic, biologic, social, etc.) at differing scales and interpreted by several disciplines due to the interdisciplinary nature of the problem. The research team achieves this integration through the use of a "collaboratory," that is, a web-based system where different researchers and users can come together to build a system of data, predictive models, and management projects. A collaboratory develops and evolves over time, adapting to new data and modeling systems and the needs of the users of the systems projects, such as seasonal forecasts of malaria intensity in the case study.

Another possible model would mix support for observatories and test beds with programs with shorter timescales and Integrative Graduate Education and Research Traineeships (IGERTs); a third would be an analogy to the NSF Engineering Research Centers, which have a 10-year time horizon. All of these funding models would have as a goal the field demonstration of integrated sensors (across variables and platforms) further integrated with cyberinfrastructure development bridging the critical gap of single sensor demonstration to integrated field demonstration.

NASA Airborne and Spaceborne Sensor Technology Research and Development

As noted earlier, a review of, and recommendations concerning, the development of new airborne and spaceborne sensors by NASA, known as the "decadal survey" (NRC, 2007) has recently been completed. The recommendations of the survey for specific water-related missions are summarized in Appendix C. However, another important recommendation of the decadal survey was that "NASA should support Earth science research via suborbital platforms: airborne programs, which have suffered substantial diminution, should be restored. . ."

As shown in the Arctic case study in Chapter 4, certain remotely sensed measurements are more useful from airborne platforms than from spaceborne platforms, and the capability of low-cost airborne remote sensing needs to be further developed and integrated with other measurement systems.

In recommendation 2-1, the report discusses the interagency gaps from sensor development to operational deployment. This recommendation is analogous but focused on NASA. NASA is unique in that it has research and development programs that span from sensor development to operational space deployment, often with other agencies like NOAA or the Department of Defense. Historically NASA demonstrated sensors through a strong airborne program, carried out "integrated field demonstrations" through validation campaigns related to Earth Observing System (EOS) missions and Earth System Science Pathfinder (ESSP) exploratory missions that often included in-situ and airborne measurements, and supported research and application in the use of space observations.

Unfortunately, the airborne remote sensing component within NASA has been weakened for both piloted and unpiloted systems, causing a gap in the sensor development-to-operational deployment road. This weakness is being felt in the sensor development activities because there is no obvious mechanism for easily and inexpensively testing new sensor concepts. This weakness may also extend to operational deployments because the possibility that new sensors, or the operational products from new sensor measurements, may be best delivered through airborne systems rather than through spaceborne systems that cannot be evaluated effectively.

Recommendation 3-1: *NASA should strengthen its program in sensor technology research and development, including piloted and unpiloted airborne sensor deployment for testing new sensors and as a platform for collecting and transmitting data useful for applications.*

In the current federal budget environment NASA has constrained resources. The space agencies in other countries have similar pressures. Historically there have been joint programs between NASA and international space agencies in developing and launching specific sensors. Two examples are the Tropical Rainfall Measurement Mission (TRMM) with the Japanese Aerospace Exploration Agency (JAXA) and the Calipso satellite with the French space agency, Centre National d'Études Spatiales (CNES).

Regardless of the above collaborations, NASA can do more, particularly in the areas of hydrologic and environmental sciences, to foster both U.S. interagency and international partnerships.

Recommendation 3-2: *In addition to partnerships with other federal agencies for the development and testing of experimental sensors that are of a particular interest to agencies, the Nation, and especially NASA, should explore additional strategic partnerships with space agencies in other countries and regions, such as the European Space Agency (ESA), JAXA, CNES, and the Canadian Space Agency (CSA).*

Early Integration of Satellite and Airborne Sensing into Observatory Design

Assuming a strengthening of NASA's sensor development programs as recommended above, there is an opportunity for NASA to take a leadership role in developing integrated ground-to-space sensor networks as envisioned in Chapters 2 and 3 of this report. Some of the necessary coordination could be done through the Office of Science and Technology Policy's Subcommittee on Water Availability and Quality (SWAQ), as proposed by the NRC (2006c), but this should be only one of many points of contact. This is an area in which an increase in support by NASA could be highly leveraged by other agencies. The decadal survey (NRC, 2007) recommended that NASA increase support of its Research and Analysis (R&A) program, noting that among other purposes the R&A program is "necessary for improving calibrations and evaluating the limits of both remote and in situ data." Engagement in observatory design would also be consistent with NASA's Earth Science Implementation Plan for Energy and Water Cycle Research (NASA, 2007), which notes that "[i]n some cases, NASA investments may be required to supplement [planned] activities to ensure that they meet specific needs, for example, in situ measurements of parameters that are essential to validating space based remote sensing, as well as quantities needed but not otherwise measured or derived."

Recommendation 4-1: *NASA and NOAA should work with NSF and other agencies to assure that plans for incorporation of space-based and airborne observations (from both existing and, preferably, planned or proposed missions) are part and parcel of the experimental design of these proposed observatories.*

Advanced Cyberinfrastructure

Cyberinfrastructure is a key linkage between observatories, and a way to achieve economies of scale within and between multidisciplinary initiatives. Yet there is insufficient work in exploiting current advances in cyberinfrastructure. The underfunding of cyberinfrastructure has been shown to compromise results and data interpretation at all scales. Advances in computational, communication, and Internet technologies will help develop and deliver "services" to the broad community of users.

The observatories and other national initiatives offer settings and mechanisms for the development and testing of the "next generation" service delivery system. For example, current NOAA plans for the newly funded National Integrated Drought Information System (NIDIS) include the development of a "drought portal" for the delivery of drought information.

The guiding principles important to a next generation system were set forth in Chapter 3. These included employing open architecture solutions to enable the rapid adoption of new hardware and software technologies and nonproprietary and, ideally, open-source software solutions (e.g., middleware, metadata management protocols) and promote the modularity, extensibility, scalability, and security that are needed for observation and forecasting. They also included using community standards (e.g., data transport, quality assurance/quality control, metadata specifications, interface operations) to ensure system interoperability, and providing open access to data and information through customizable portals to ensure timely access to data, information, and forecasts.

Recommendation 5-1: *Advanced cyberinfrastructure should not only be incorporated as part of planned observatories and related initiatives to help manage, understand, and use diverse data sets, but should be a central component in their planning and design.*

Recommendation 5-2: *Utilization of web-based services, such as collaboratories, for the distribution of observations, model predictions, and related products to potential users, should be encouraged.*

To achieve efficiencies and economy of scale, it may be useful to create an interobservatory council among at least the NEON and WATERS communities to address interoperability issues and agree on a uniform cyberinfrastructure. An important step forward has been taken by WATERS, in the form of a recently funded "Test-bed Digital Observatory" for the Susquehanna River Basin and Chesapeake Bay, which will attempt to demonstrate the applicability and utility of many aspects of the Hydrologic Information System (HIS) from CUAHSI and the "CyberCollaboratory" from Collaborative Large-Scale Engineering Analysis Network for Environmental Research (CLEANER) for use by both the hydrologic and water-quality communities.

The NRC (2006c) went one step beyond an interobservatory council and recommended that "[s]erious consideration should be given to placing the various NSF environmental observatory programs under a parent organization that could be termed something like the 'Environmental Observatory Networks' or EON.' Such a parent entity would be responsible for cyberinfrastructure development, among other shared activities." This approach also seems reasonable.

Water and environmental data products from the observatories can be extremely useful for the classrooms of the nation. An exploratory effort to do this—the Student Analysis of Data Driving Learning about the Earth (SADDLE) project—is creating student-friendly interfaces to allow students to interact with real-time and archived databases such as would be associated with a large-scale scientific observatory. There is a need for various studies of this kind to understand how observatory-derived data can be delivered and used in classrooms at various educational levels. It should be noted that the parent organization EON, proposed by the NRC (2006c), would also have a role in coordinating educational outreach by the various proposed NSF environmental observatories.

Data Assimilation and Demonstration Projects

As discussed in a number of case studies, there is an unmet need to have sensor systems that link plot-scale, in-situ measurements to measurements at satellite scales, and the related data-assimilation systems that merge these data with appropriate predictive models. It is recognized that both NASA and NSF have research programs that support data assimilation. However, these ongoing programs are not sufficient.

Recommendation 6-1: *NASA and NSF should develop and strengthen program elements focused on demonstration projects and application of data assimilation in operational settings where researchers work collaboratively with operational agencies.*

This recommendation is consistent with the decadal survey (NRC, 2007) recommendation that "NASA, NOAA, and USGS should increase their support for Earth system modeling, including provision of high-performance computing facilities and support for scientists working in the areas of modeling and data assimilation." It is also consistent with NASA's implementation plan for water and energy cycle research (NASA, 2007), which includes "Creating a global land and atmosphere data assimilation system for energy and water variables" among its major elements. Existing data sets should be fully exploited in these efforts.

Remote Sensing: Operational Systems and Data Products

Many of the above recommendations have been especially applicable to observatories, experimental watersheds, and other kinds of fairly focused work. But it is important to look beyond observatories to other, broader, downstream applications of observations.

Chapter 2 makes clear that the Nation has invested significantly in remote sensing, especially in space observations. However, agencies such as NASA and NOAA need to foster activities that will help state agencies and other federal agencies utilize and merge remote sensing measurements with process-based hydrological models for improved management decisions. The availability through the NASA Distributed Active Archive Centers (DAACs) of data from the recent suite of Earth Observations System satellite sensors is significant, and Advanced Very High Resolution Radiometer (AVHRR) and Landsat Thematic Mapper (TM) data are widely used. Yet the case studies in Chapter 4, and in other reports, indicate that remote sensing data and data products are often underutilized in operational and application uses. NASA and NOAA have often encountered difficulties in "Crossing the Valley of Death" (NRC, 2000b) from research to operations. This has often resulted in underutilization of data and data products, resulting in long delays in users and agencies realizing the value of such measurements to society.

Also, the process for identifying the remotely sensed measurement needs of the community and federal agencies, and for the subsequent technological development of such sensors, is often unclear. One example of need is water-quality measurements from space, where only the most basic water-quality variables, turbidity and chlorophyll, have been measured and only as an experimental variable. The scientific basis for water-quality measurements from space needs considerable development, and it does not appear that any NASA program is addressing this problem.

Integrated data products that merge satellite, airborne, and ground observations with model and decision-support products at regional to continental scales, would be of particular value. Demonstration projects at observatories, including remote sensing validation observatories, may be an effective mechanism to demonstrate to users of water cycle and environmental data products how they may better exploit the current investment in space observations in their prediction and management activities.

Recommendation 7-1: *NASA should take the lead by expanding support for the application of integrated satellite remote sensing data products. NSF, NOAA, and other federal and state agencies engaged in environmental sensing should likewise expand support for the creation of the integrated digital products that meet educational, modeling, and decision-support needs.*

Utilization of remote sensing data in applications and operational settings also depends on assurance by the government to users that these measurements will be maintained into the future so their investment in developing procedures that use the data will be realized. Except for atmospheric remote sensing, which NASA develops for NOAA and Department of Defense to use operationally, and visible wavelength surface imagers (AVHRR and TM), all other NASA-developed sensors are viewed as limited-duration experimental sensors. Transitioning sensors from a research to an operational monitoring mode is critical if users and agencies are to utilize their products as part of their management responsibilities.

However, as yet there is no government strategy for such a transition beyond those sensors used for weather. Current concerns about the elimination on the new NASA Landsat satellite of the infrared (IR) channels, which are used by USDA and state agencies (among others) to monitor irrigation water usage, is a case in point. The elimination is due to budgetary issues and lack of consensus over who will pay for the sensor. Similar issues are occurring on the planned National Polar Orbiter Environmental Satellite System (NPOESS). This is one important impediment for wider usage of remote sensing measurements in applications and by operational agencies.

***Recommendation** 7-2: Congress, through the budgetary process, should develop a strategy for transitioning NASA experimental satellite sensors to operational systems with assured data continuity so that the Nation's investment in remote sensing can be utilized over the long term by other federal agencies and users.*

Water Agencies

A wide variety of water agencies have a need for accurate and timely information from in-situ, airborne, and spaceborne sensors. These agencies may be involved in anything from water supply forecasting to flood forecasting to transportation and recreation. They are often underfunded, with a very modest number of trained scientists and engineers. Because of this, once they develop a system—even an imperfect one—that provides reasonably good information on which to make their decisions, there is strong aversion to new kinds of data, models, and tools that have less of a track record.

Adopting new systems involves start-up costs in the way of model design and calibration, and staff training. Users who are adapted to a given product will likely need to be educated in interpretation and use of the new product in order to assist them through the transitional period. Further, until a new remote sensing system is fully operational with a firm commitment of continued service from the federal agency responsible for it, a local or state water agency may be very conservative about adopting the new technology.

Nonetheless, there are numerous opportunities for both existing and new sensor data and information to be incorporated into planning and forecasting, and such opportunities will only increase in the future.

Recommendation 8-1: *Water agencies should be alert for opportunities to incorporate new sensor and modeling technologies that will allow them to better deliver their mission.*

References

Ahn, J., S. B. Grant, C. Q. Surbeck, P. M. DiGiacomo, and N. Nezlin. 2005. Coastal water quality impact of storm runoff from and urban watershed in southern California. Environmental Science and Technology 39:5490-5953.

Alley, W. M., R. W. Healy, J. W. LaBaugh, and T. E. Reilly. 2002. Flow and storage in groundwater systems. Science 296(5575):1985-1990.

Allison, G. B., P. G. Cook, S. R. Barnett, G. R. Walker, I. D. Jolly, and M. W. Hughes. 1990. Land clearance and river salinisation in the western Murray Basin, Australia. Journal of Hydrology 119:1-20.

Alsdorf, D. E., C. Birkett, T. Dunne, J. Melack, and L. Hess. 2001. Water level changes in a large Amazon lake measured with spaceborne radar interferometry and altimetry. Geophysical Research Letters 28(14):2671-2674, 10.1029/2001GL012962.

Alsdorf, D. E., and D. P. Lettenmaier. 2003. Tracking fresh water from space. Science 301:1491-1494.

Alsdorf, D. E., E. Rodriguez, and D. Lettenmaier. 2007. Measuring surface water from space. Reviews of Geophysics 45(2):RG2002, doi:10.1029/ 2006RG 000197.

Amman R. I., W. Ludwig, and K. H. Schleifer. 1995. Phylogenetic identification and in situ detection of individual cells without cultivation. Microbiol. Rev. 59:143-169.

Anagnostou, E. 2004. Overview of overland satellite rainfall estimation for hydro-meteorological applications. Surveys in Geophysics 25:511-537.

Anderson, M. C., J. M. Norman, J. R. Mecikalski, R. D. Torn, W. P. Kustas, and J. B. Basara. 2004. A multi-scale remote sensing model for disaggregating regional fluxes to micrometeorological scales. J. Hydromet. 5:343-363.

Bailey, S., and P. Werdell. 2006. A multi-sensor approach for the on-orbit validation of ocean color satellite data products. Remote Sensing of Environment 102(1-2):12-23.

Baldocchi, D., R. Valentini, S. Running, W. Oechel, and R. Dahlman. 1996.

Strategies for measuring and modeling carbon dioxide and water vapour fluxes over terrestrial ecosystems. Global Change Biology 2(3):159-168.

Bales, R., and R. Rice. 2006. Snowcover along elevation gradients in the Upper Merced River basin of the Sierra Nevada of California from MODIS and blended ground data. Eos Trans. AGU, 87(52), Fall Meet. Suppl., Abstract C31C-02.

Bales, R. C., N. P. Molotch, T. H. Painter, M. D. Dettinger, R. Rice, and J. Dozier. 2006. Mountain hydrology of the western United States. Water Resources Research W08432, doi:10.1029/2005WR004387.

Bales, R. C., K. A. Dressler, B. Iman, S. R. Fassnacht, and D. Lampkin. 2008. Fractional snow cover in the Colorado and Rio Grande basins, 1955-2002. Water Resources Research 44, W01425, doi:10.1029/2006WR005377.

Band, L. E., P. Patterson, R. R. Nemani, and S. W. Running. 1993. Forest ecosystem processes at the watershed scale: 2. Adding hillslope hydrology. Agricultural and Forest Meteorology 63:93-126.

Band, L. E., C. L. Tague, P. Groffman, and K. Belt. 2001. Forest ecosystem processes at the watershed scale: Hydrological and ecological controls of nitrogen export. Hydrological Processes 15:2013-2028.

Barrenetxea, G., O. Couach, M. Krichane, T. Varidel, S. Mortier, J. Mezzo, M. Bystranowski, S. Dufey, H. Dubois-Ferrière, J. Selker, M. Parlange, and M. Vetterli. 2006. SensorScope: An Environmental Monitoring Network. Eos Trans. AGU, 87(52), Fall Meet. Suppl., Abstract H51D-0513.

Basseville, M., A. Benveniste, K. C. Chou, S. A. Golden, R. Nikoukhah, and A. S. Willsky. 1992. Modeling and estimating multiresolution stochastic processes. IEEE Trans. On Inform. Theory 38(2):766-784.

Bastiaanssen, W. G. M., M. Menenti, R. A. Feddes, and A. A. M. Holtslag. 1998a. A remote sensing surface energy balance algorithm for land (SEBAL) 1. formulation. J. Hydrol. 212-213:198-212.

Bastiaanssen, W. G. M., H. Pelgrum, J. S. Y. Wang, Y. Ma, J. F. Moreno, G. K. Roerink, and T. van der Wal. 1998b. A remote sensing surface energy balance algorithm for land (SEBAL) 2. validation. Journal of Hydrology 212-213:213-229.

Bawden, G. W., W. Thatcher, R. S. Stein, K. W. Hudnut, and G. Peltzer. 2001. Tectonic contraction across Los Angeles after removal of groundwater pumping effects. Nature 412:812-815.

Bendikov, T. A., J. Kim, and T. C. Harmon. 2005. Development and environmental application of a nitrate selective microsensor based on doped polypyrrole films. Sensors and Actuators B: Chemical 106(2):512-517.

Bennett, A. F. 1993. Inverse Methods in Physical Oceanography. New York: Cambridge University Press. 346 pp.

Bennett, P. C., F. K. Hiebert, and W. J. Choi. 1996. Microbial colonization and weathering of silicates in a petroleum-contaminated groundwater. Chemical Geology 132:45-53.

Birkett, C. 1995. The contribution of TOPEX/POSEIDON to the global monitoring of climatically sensitive lakes. J. Geophys. Res. 100(C12): 25179-25204.

Birkett, C. 1998. Contribution of the TOPEX NASA radar altimeter to the global monitoring of large rivers and wetlands. Water Resources Research 34 (5):1223-1240.

Birkett, C. M., L. A. K. Mertes, T. Dunne, M. H. Costa, and M. J. Jasinski. 2002. Surface water dynamics in the Amazon Basin: Application of satellite radar altimetry. J. Geophysical Res. 107(D20, 8059):26.1-26.21, doi:10.1029/2001 JD000609.

Bruinsma, J., ed. 2003. World Agriculture: Towards 2015/2030, an FAO Study. London, England: Earthscan Publications.

Buck, S. M., H. Xu, M. Brasuel, M. A. Philbert, and R. Kopelman. 2004. Nanoscale probles encapsulated by biologically localized embedding (PEBBLEs) for ion sensing and imaging in live cells. Molecular Recognition and Chemical Sensor 63(1):41-59.

Buffle, J. and Horvai, G. 2000. In-situ Monitoring of Aquatic Systems: Chemical Analysis and Speciation. New York, NY: Wiley.

Burkholder, J., D. Eggleston, H. Glasgow, C. Brownie, R. Reed, G. Janowitz, M. Burkholder, J. M., M. A. Mallin, H. B. Glasgow Jr., L. M. Larsen, M. R. McIver, G. C. Shank, N. Deamer-Melia, D. S. Briley, J. Springer, B. W. Touchette, and E. K. Hannon. 1997. Impacts to a coastal river and estuary from rupture of a large swine waste holding lagoon. Journal of Environmental Quality 26:1451-1466.

Posey, G. Melia, C. Kinder, R. Corbett, D. Toms, T. Alphin, N. Deamer, and J. Springer. 2004. Comparative impacts of two major hurricane seasons on the Neuse River and western Pamlico Sound ecosystems. Proceedings of the National Academy of Sciences 101:9291-9296.

California Bay-Delta Authority. 2000. Final Programmatic Environmental Impact Statement/Environmental Impact Report. Accessed on-line November 30, 2006, at http://calwater.ca.gov/CALFEDDocuments /Final_EIS_ EIR.shtml.

Carpenter, T. M., J. A. Sperfslage, K. P. Georgakakos, T. Sweeney, and D. L. Fread. 1999. National threshold runoff estimation utilizing GIS in support of operational flash flood warning systems. Journal of Hydrology 224:21-44.

Cayan, D., M. VanScoy, M. Dettinger, and J. Helly. 2003. The wireless watershed in Santa Margarita Ecological Reserve. Southwest Hydrology 2:18-19.

Chapin, F. S., III, M. Sturm, M. C. Serreze, J. P. McFadden, J. R. Key, A. H. Lloyd, A. D. McGuire, T. S. Rupp, A. H. Lynch, J. P. Schimel, J. Beringer, W. L. Chapman, H. E. Epstein, E. S. Euskirchen, L. D. Hinzman, G. Jia, C.-L. Ping, K. D. Tape, C. D. C. Thompson, D. A. Walker, and J. M. Welker. 2005. Role of land-surface changes in arctic summer warming. Science 310:657-660.

Cline, D. W., R. C. Bales, and J. Dozier. 1998. Estimating the spatial distribution of snow in mountain basins using remote sensing and energy balance modeling. Water Resources Research 34(5):1275-1285, doi:10.1029/97WR03755.

Collins, P. G., K. Bradley, M. Ishigami, and A. Zettl. 2000. Extreme oxygen sensitivity of electronic properties of carbon nanotubes. Science 287(5459): 1801-1804.

Conklin, M., R. Bales, E. Boyer, D. Cayan, J. Dozier, G. Fogg, T. Harmon, J. Kirchner, N. Miller, N. Molotch, and K. Redmond. 2006. Observatory design in the mountain west: scaling measurements and modeling in the San Joaquin Valley and Sierra Nevada. Eos Trans. AGU 87(52), Fall Meet. Suppl., Abstract H21F-1430.

Costanza, R., and A. Voinov, eds. 2004. Landscape Simulation Modeling: A Spatially Explicit, Dynamic Approach. New York, NY: Springer-Verlag.

Cotofana, C., L. Ding, P. Shin, S. Tilak, T. Fountain, J. Eakins, and F. Vernon. 2006. An SOA-based framework for instrument management for large-scale observing systems (USArray case study). Pp. 815-822 in Institute of Electrical and Electronics Engineers International Conference on Web Services (ICWS'06). Washington, D.C.: IEEE Computer Society.

Crossett, K. M., T. J. Culliton, P. C. Wiley, and T. R. Goodspeed. 2004. Population trends along the coastal United States, 1980-2008. National Oceanic and Atmospheric Administration, NOAA's National Ocean Service, Management and Budget Office, Special Projects. 54 pp.

Crow, W. T., and E. F. Wood. 2003. The assimilation of remotely sensed soil brightness temperature imagery into a land surface model using Ensemble Kalman filtering: A case study based on ESTAR measurements during SGP97. Advances in Water Resources 26:137-149.

Cui, Y., Q. Wei, H. Park, and C. M. Liebler. 2001. Nanowire nanosensors for highly sensitive and selective detection of biological and chemical species. Science 293(5533):1289-1292.

Dalton, J. B., D. J. Bove, C. S. Mladinich, and B. W. Rockwell. 2004. Identification of spectrally similar materials using the USGS Tetracorder algorithm: the calcite-epidote-chlorite problem. Remote Sensing of the Environment 89: 455-466.

Daniel, M. M., and A. S. Willsky. 1997. A multiresolution methodology for signal-level fusion and data assimilation with applications to remote sensing. Proceedings of the IEEE 85(1):164-180.

Delin, G. N., R. W. Healy, M. K. Landon, and J. K. Bohlke. 2000. Effects of topography and soil properties on recharge at two sites in an agricultural field. Journal of the American Water Resources Association 36(6):1401-1416.

Delin, K. A. 2002. The sensor web: A macro-instrument for coordinated sensing. Sensors 2:270-285.

Delin, K. A., S. P. Jackson, D. W. Johnson, S. C. Burleigh, R. R.Woodrow, J. M. McAuley, J. M. Dohm, F. Ip, T. P. A. Ferré, D. F. Rucker, and V. R. Baker. 2005. Environmental studies with the sensor web: principles and practice. Sensors 5:103-117.

Dennehy, K. F. 2000. High Plains regional ground-water study. U.S. Geological Survey Fact Sheet FS-091-00. 6 pp.

Dhariwal, A., B. Zhang, C. Oberg, B. Stauffer, A. Requicha, D. Caron, and G. S. Sukhatme. 2006. Networked aquatic microbial observing system. IEEE International Conference on Robotics and Automation, Orlando, Fl. Pp. 4285-4287.

Dozier, J., and T. H. Painter. 2004. Multispectral and hyperspectral remote sensing of alpine snow properties. Ann. Rev. Earth Planet. Sci. 32:465-494, doi:10.1146/annurev.earth. 32.101802.120404.

Duval, J. S. 1977. High sensitivity gamma-ray spectrometry-state of the art and trail application of factor analysis. Geophysics 42:549-559.

Earman, S., A. R. Campbell, F. M. Phillips, B. D. Newman. 2006. Isotopic exchange between snow and atmospheric water vapor: Estimation of the snowmelt component of groundwater recharge in the southwestern United States. Journal of Geophysical Research 111, D09302, doi:10.1029/2005 JD006470.

Edmunds, W. M., and C. B. Gaye. 1997. Naturally high nitrate concentrations in groundwater from the Sahel. Journal of Environmental Quality 26:1231-1239.

England, A. W., and R. D. De Roo. 2006. Active Layer Thickness and Moisture Content of Arctic Tundra from SVAT/Radiobrightness Models and Assimilated 1.4 or 6.9 GHz Brightness. Final report of NSF Award ID 0240747. Ann Arbor, Mich.: College of Engineering, University of Michigan.

Entekhabi, D., H. Nakamura, and E. G. Njoku. 1994. Solving the inverse problem for soil moisture and temperature profiles by sequential assimilation of multifrequency remotely sensed observations. IEEE Transactions on Geoscience and Remote Sensing 32(2):438-448.

Epstein, P. R., H. F. Diaz, S. Elias, G. Grabherr, N. E. Graham, W. J. M. Martens, E. Mosley-Thompson, and J. Susskind. 1998. J. Biological and Physical Signs of Climate Change: Focus on Mosquito-Borne Diseases. Bulletin of the American Meteorological Society 79:409-417.

Estrin, D., W. Michener, and G. Bonito. 2003. Environmental Cyberinfrastructure Needs for Distributed Sensor Networks: A Report from a National Science Foundation Sponsored Workshop. Scripps Institute of Oceanography. Available on-line at http://www.lternet.edu/sensor_ report.

Evensen, G. 1994. Sequential data assimilation with a nonlinear quasi-geostrophic model using Monte Carlo methods to forecast error statistics. Journal of Geophysical Research 99(C5):10143-10162.

Fahlquist, L. 2003. Ground-water quality of the Southern High Plains aquifer, Texas and New Mexico, 2001. Open-File Rept. 03-45. 59 pp.

Famiglietti, J. S., and E. F. Wood. 1994. Multi-scale modeling of spatially-variable water and energy balance processes. Water Resources Research 30(11): 3061-3078.

Famiglietti, J. S. 2004. Remote sensing of terrestrial water storage, soil moisture and surface waters, in the state of the planet: Frontiers and challenges. J. Sparks and C. J. Hawkesworth, eds. Geophysics, Geophysical Monograph Series 150:197-207.

Fassnacht, S. R., K. A. Dressler, and R. C. Bales. 2003. Snow water equivalent interpolation for the Colorado River Basin from snow telemetry (SNOTEL) data. Water Resources Research 39(8):1208, doi: 10.1029/2002WR00 1512.

Finnerty, B. D., M. B. Smith, D. J. Seo, V. Koren, and G. E. Moglen. 1997. Space-time scale sensitivity of the Sacramento model to radar-gage precipitation inputs. Journal of Hydrology 203:21-38.

Fischman, M. A., and A. W. England. 1999. Sensitivity of a 1.4 GHz Direct-sampling Digital Radiometer. IEEE Trans. Geosci. Remote Sensing 37: 2172–2180.

Fischman, M. A., A. W. England, and C. S. Ruf. 2002. How digital correlation affects the fringe washing function in L-Band aperture synthesis radiometry. IEEE Trans. Geosci. Remote Sensing 40:671-679.

Fread, D. L., R. C. Shedd, G. F. Smith, R. Farnsworth, C. N. Hoffeditz, L. A. Wenzel, S. M. Wiele, J. A. Smith, and G. N. Day. 1995. Modernization in the National Weather Service river and flood program. Weather and Forecasting 10(3):477-484.

Fritzsche, A. E., and Z. G. Burson. 1973. Airborne gamma radiation surveys for snow water-equivalent research: Progress report. Technical Report No. EGG-1183-1623, OSTI ID: 4259276. 54 pp.

Galloway, D. L., K. W. Judnut, S. E. Ingebritsen, S. P. Phillips, G. Peltzer, F. Rogez, and P. A. Rosen. 1998. Detection of aquifer system compaction and land subsidence using interferometric synthetic aperture radar, Antelope Valley, Mojave Desert, California. Water Resources Research 34 (1):2573-2585.

Georgakakos, K. P. 2006. Analytical results for operational flash flood guidance. Journal of Hydrology 317(1-2):81-103.

Georgakakos, K. P., N. E. Graham, T. M. Carpenter, A. P. Georgakakos, and H. Yao. 2005. Integrating climate-hydrology forecasts and multi-objective reservoir management for Northern California. EOS 86(12):122-127.

Gilbride, K. A., D.-Y. Lee, and L. A. Beaudette. 2006. Molecular techniques in wastewater: Understanding microbial communities, detecting pathogens, and real-time process control. Journal of Microbiological Methods 66:1-20.

Goeders, K. M., J. S. Colton, and L. A. Bottomley. 2008. Microcantilevers: Sensing Chemical Interactions via Mechanical Motion. Chem. Rev. 108(2):522-542.

Goodkind, J. M. 1999. The superconducting gravimeter. Rev. Sci. Instruments 70:4131-4152.

Gooseff, M. N., D. M. McKnight, R. L. Runkel, and J. H. Duff. 2004. Denitrification and hydrologic transient storage in a glacial meltwater stream, McMurdo Dry Valleys, Antarctica. Limnology and Oceanography 49(5): 1884-1895.

Gorenburg I. P., D. McLaughlin, and D. Entekhabi. 2001. Scale-recursive assimilation of precipitation data. Advances in Water Resources 24(9-10):941-953.

Haack, E. A., and L. A. Warren. 2003. Biofilm hydrous manganese oxides and metal dynamics in acid rock drainage. Environmental Science and Technology 37:4138-4147.

Hall, D. K., G. A. Riggs, V. A. Salomonson, N. E. DiGirolamo, and K. J. Bayr. 2002. MODIS snow cover products. Remote Sensing of Environment 83:181-194.

Hamilton, M. P., E. A. Graham, P. W. Rundel, M. F. Allen, W. Kaisser, M. H. Hansen, and D. L. Estrin. 2007. New approaches in embedded networked sensing for terrestrial ecological observatories. Environmental Engineering Science 24(2):192-204.

Hamilton, S. K., J. L. Tank, D. F. Raikow, W. M. Wollheim, B. J. Peterson, and J. R. Webster. 2001. Nitrogen uptake and transformation in a Midwestern U.S. stream: A stable isotope enrichment study. Biogeochemistry 54(3):297-340.

Hanson, P. C., D. L. Bade, S. R. Carpenter, and T. K. Kratz. 2003. Lake metabolism: Relationships with dissolved organic carbon and phosphorus. Limnology and Oceanography 48:1112-1119.

Harmon, T., R. Ambrose, R. Gilbert, J. Fisher, and M. Stealey. 2007. High-resolution river hydraulic and water quality characteristics using rapidly deployable networked infomechanical systems (NIMS RD). Environmental Engineering Science 24(2):151-159.

Hatchett, D. W., and M. Josowicz. 2008. Composites of Intrinsically Conducting Polymers as Sensing Nanomaterials. Chem. Rev. 108(2):746-769.

Hay, S. I., C. J. Tucker, D. J. Rogers, and M. J. Packer. 1996. Remotely sensed surrogates of meteorological data for the study of the distribution and abundance of arthropod vectors of disease. Ann. Trop. Med. Parasitol. 90:1-19.

Hay, S. I., J. Cox, D. J. Rogers, S. E. Randolph, D. I. Stern, G. D. Shanks, M. F. Myers, and R. W. Snow. 2002. Climate change and the resurgence of malaria in the East African Highlands. Nature 415:905-909.

Hayashi, M., G. van der Kamp, and R. Schmidt. 2003. Focused infiltration of snowmelt water in partially frozen soil under small depressions. Journal of Hydrology 270:214-229.

Hoffman, J., H. A. Zebker, D. L. Galloway, and F. Amelung. 2001. Seasonal subsidence and rebound in Las Vegas Valley, Nevada, observed by synthetic aperture radar interferometry. Water Resources Research 37(6):1551-1566.

Houser, P. R., W. J. Shuttleworth, H. V. Gupta, J. S. Famiglietti, K. H. Syed, and D. C. Goodrich. 1998. Integration of soil moisture remote sensing and hydrologic modeling using data assimilation. Water Resources Research 34(12):3405-3420.

Howell, T. A., S. R. Evett, J. A. Tolk, and A. D. Schneider. 2004. Evapotranspiration of full-deficit-irrigated, and dryland cotton on the northern Texas high plains. J. Irrigation and Drainage Engineering-ASCE 130:277-285.

Huffman, G. J., R. F. Adler, P. Arkin, A. Chang, R. Ferraro, A. Gruber, J. Janowiak, A. McNab, B. Rudolf, and U. Schneider. 1997. The Global Precipitation Climatology Project (GPCP) Combined Precipitation Dataset. Bull. Amer. Meteorol. Soc. 78:5-20.

Hutson, S. S., N. L. Barber, J. F. Kenny, K. S. Linsey, D. S. Lumia, and M. A. Maupin. 2004. Estimated use of water in the United States in 2000. U.S. Geological Survey Circular 1268. Available on-line at http://water.usgs.gov/pubs/circ/ 2004/circ1268/. Reston, Va.: U.S. Geological Survey.

Hyyppa, J., and M. Hallikainen. 1993. A helicopter-borne eight-channel ranging scatterometer for remote sensing. II-forest inventory. IEEE Transactions on Geoscience and Remote Sensing 31:170-179.

Inoue, Y., S. Morinaga, and A. Tomita. 2000. A blimp-based remote sensing system for low-altitude monitoring plant variables: A preliminary experiment for agricultural and ecological applications. Int. J. Remote Sensing 21:379-385.

Ivanov V. Y., E. R. Vivoni, R. L. Bras, and D. Entekhabi. 2004a. Catchment hydrologic response with a fully-distributed triangulated irregular network model. Water Resources Research 40:W11102.

Ivanov, V. Y., E. R. Vivoni, R.L. Bras, and D. Entekhabi. 2004b. Preserving high-resolution surface and rainfall data in operational-scale basin hydrology: A fully-distributed physically-based approach. Journal of Hydrology 298:1-4.

Jackson, R. D. 1984. Remote sensing of vegetation characteristics for farm management. Remote Sens. 475:81-96.

Jackson, T., and P. O'Neill. 1987. Temporal observations of surface soil moisture using a passive microwave sensor. Remote Sens. Env. 21:281-296.

Jackson, T. J., D. LeVine, C. Swift, T. Schmugge, and F. Schiebe. 1995. Large area mapping of soil moisture using the ESTAR passive microwave radiometer in Washita'92. Rem. Sens. Env. 53: 27-37.

Jackson, T. J., D. M. LeVine, A. Y. Hsu, A. Oldak, P. J. Starks, C. T. Swift, J. D. Isham and M. Haken. 1999. Soil moisture mapping at regional scales using microwave radiometry: The Southern Great Plains Hydrology Experiment. IEEE Transactions on Geoscience and Remote Sensing 37(5):2136-2151.

Jackson, T. J., A. J. Gasiewski, A. Oldak, M. Klein, E. G. Njoku, A. Yevgrafov, S. Christiani, and R. Bindlish. 2002. Soil moisture retrieval using the C-band polarimetric scanning radiometer during the Southern Great Plains

1999 Experiment. IEEE Transactions on Geoscience and Remote Sensing 40(10):2151-2161.

Jacobs, J., W. Krajewski, H. Loescher, R. Mason, K. McGuire, B. Mohanty, G. Poulos, P. Reed, J. Shanley, O. Wendroth, and D. A. Robinson. 2006. Enhanced Water Cycle Measurements for Watershed Hydrologic Sciences Research. A Report to the Consortium of Universities for the Advancement of Hydrologic Sciences, Inc. Accessed on-line at http://www.cuahsi.org/hmf/docs/watercycle-20060517-draft.pdf.

Jeong, Y., B. F. Sanders, and S. B. Grant. 2006. The information content of high frequency environmental monitoring data signals pollution events in the coastal ocean. Environmental Science and Technology 40(20): 6215-6220.

Jianrong, C., M. Yuqing, H. Nongyue, W. Ziaohua, and L. Sijiao. 2004. Nanotechnology and biosensors. Biotechnology Advances 22(7):505-518.

Jonkman, S. N. 2005. Global perspectives on loss of human life caused by floods. Natural Hazards 34(2):151-175.

Junkermann, W. 2000. An ultralight aircraft as platform for research in the lower troposphere: System performance and first results from radiation transfer studies in stratiform aerosol layers and broken cloud conditions. J. Atmospheric and Oceanic Tech. 18:934-946.

Kelly, E. J., A. T. C. Chang, J. L. Foster, and D. K. Hall. 2004. Using remote sensing and spatial models to monitor snow depth and snow water equivalent. PP. 35 to 59 in Spatial Modelling of the Terrestrial Environment. R. Kelly, N. Drake, and S. Barr, eds. Hoboken, New Jersey: John Wiley & Sons, Ltd.

Kendall, C. 1998. Tracing nitrogen sources and cycles in catchments. Pp. 519-576 in Isotope Traces in Catchment Hydrology. C. Kendall and J. J. McDonnell, eds. Amsterdam: Elsevier.

Kerr, Y. H., P. Waldteufel, J. P. Wigneron, J. M. Martinuzzi, J. Font, and M. Berger. 2001. Soil moisture retrieval from space: The Soil Moisture and Ocean Salinity Mission (SMOS). IEEE. Trans. Geosci. Remote Sens. 39: 1729-1735.

Kim, E. J., J. C. Shiue, T. Doiron, C. Principe, and A. Rodriguezs. 2000. A preview of AMSR: Airborne C-band microwave radiometer (ACMR) observations from SGP99. Pp. 1060-1062 in Geoscience and Remote Sensing Symposium Proceedings. IGARSS 2000. IEEE 2000 Int.

Kneebone, P. E., and J. G. Hering. 2000. The behavior of arsenic and other redox-sensitive elements in Crowley Lake, CA, a reservoir in the Los Angeles aqueduct system. Environ. Sci. Technol. 34:4307-4312.

Knowles N., M. D. Dettinger, and D. R. Cayan. 2006. Trends in snowfall versus rainfall in the Western United States. Journal of Climate 19(18):4545-4559.

Lettenmaier, D. P., and J. S. Famiglietti. 2006. Water from on high. Nature 444:562-563.

LeVine, D. M., M. Kao, A. B. Tanner, C. T. Swift, and A. Griffis. 1990. Initial results in the development of a synthetic aperture microwave radiometer. IEEE Transactions on Geoscience and Remote Sensing 28:614-619.

Logan, W. S., and Rudolph, D. L. 1997. Microdepression-focused recharge in a coastal wetland, La Plata, Argentina. Journal of Hydrology 194:221-238.

Lower, B. H., R. Yongsunthon, F. P. Vellano, and S. K. Lower. 2005. Simultaneous force and fluorescence measurements of a protein that forms a bond between a living bacterium and a solid surface. Journal of Bacteriology 187:2127-2137.

Lower, S. K., M. F. Hochella, and T. J. Beveridge. 2001. Bacterial recognition of mineral surfaces: Nanoscale interactions between Shewanella and alpha-FeOOH. Science 292:1360-1363.

Lu, Z., and W. R. Danskin. 2001. InSAR analysis of natural recharge to define structure of a ground-water basin, San Bernardino, California. Geophysical Research Letters 28(13):2661-2664.

Luettgen, M. R., and A. S. Willsky. 1995. Multiscale smoothing error models. IEEE Transactions on Automatic Control 40(1):173-175.

Lunetta, R. S., J. Ediriwickrema, J. Liames, D. M. Johnson, J. G. Lyon, A. McKerrow, and A. Pilant. 2003. A quantitative assessment of a combined spectral and GIS rule-based landcover classification in the Neuse River Basin of North Carolina. Photogrammetric Engineering & Remote Sensing 69(3):299-310.

Maidment, D. R. 2004. New tools for applying GIS in water resources. Australian Journal of Water Resources 8(1): 83-92.

Maidment, D. R., J. Goodall, and G. Strassberg. 2005. Hydrology Flux, Flow and Storage. CUAHSI Hydrologic Information Systems Symposium. Austin, Tex.: University of Texas.

Margulis, S. A., D. McLaughlin, D. Entekhabi, and S. Dunne. 2002. Land data assimilation and estimation of soil moisture using measurements from the Southern Great Plains 1997 field experiment. Water Resources Research 38(12):1299, doi:10.1029/2001WR001114.

Maurice, P. A. and T. Harmon. 2007. Environmental embedded sensor networks. Environmental Engineering Science 24(2):149-150.

Maurice, P. A., D. M. McKnight, L. Leff, J. E. Fulghum, and M. Gooseff. 2002. Direct observations of aluminosilicate weathering in the hyporheic zone of an Antarctic Dry Valley stream. Geochimica et Cosmochimica Acta 66:1335-1347.

McClain, M. E., E. W. Boyer, C. L. Dent, S. E. Gergel, N. B. Grimm, P. M. Groffman, S. C. Hart, J. W. Harvey, C. A. Johnston, E. Mayorga, W. H. McDowell, and G. Pinay. 2003. Biogeochemical hot spots and hot moments at the interface of terrestrial and aquatic ecosystems. Ecosystems 6:301-312.

McCutchan, J. H. Jr., J. F. Saunders III, A. L. Pribyl, and W. M. Lewis Jr. 2003. Open channel estimation of denitrification. Limnol. Oceanogr. Methods 1:74-81.

McDonagh, C., C. S. Burke, and B. D. MacCraith. 2008. Optical Chemical Sensors. Chem. Rev. 108(2):400-422.

McGuire, V. L. 2004. Water-level changes in the High Plains aquifer, predevelopment to 2003 and 2002 to 2003. U.S. Geological Survey Fact Sheet FS-2004-3097, 6 p.

McLaughlin, D. 1995. Recent developments in hydrologic data assimilation. Reviews of Geophysics 33(Supplement):977-984.

McLaughlin, D. M. 2002. An integrated approach to hydrologic data assimilation: Interpolation, smoothing, and filtering. Advances in Water Resources 25:1275-1286.

McLaughlin, D., Y. H. Zhou, D. Entekhabi, and V. Chatdarong. 2006. Computational issues for large-scale land surface data assimilation problems. Journal of Hydrometeorology 7(3):494-510.

McMahon, P. B., K. F. Dennehy, B. W. Bruce, J. K. Bohlke, R. L. Michel, J. J. Gurdak, and D. B. Hurlbut. 2006. Storage and transit time of chemicals in thick unsaturated zones under rangeland and irrigated cropland, High Plains, United States. Water Resources Research 42, doi:10.1029/2005 WR004417.

Merwade V., G. Strassberg, J. Goodall, D. Maidment, B. Ruddell, and P. Kumar. 2005. Digital watershed for the Neuse Basin. Pp. 148-162 in Hydrologic Information System Status Report, Version 1. D. R. Maidment, ed. Available on-line at http://www.cuahsi.org/docs/ HISStatusSept15.pdf.

Michener, W.K. 2006. Meta-information concepts for ecological data management. Ecological Informatics 1:3-7.

Milly, P. C. D. 1986. Integrated remote sensing modeling of soil moisture: Sampling frequency, response time, and accuracy of estimates. Pp. 201-211 in Integrated Design of Hydrological Networks. Proceedings of the Budapest Symposium. IHAS Publication 158.

Molotch, N. P., and R. C. Bales. 2005. Scaling snow observations from the point to the grid element: Implications for observation network design. Water Resources Research 41(11): Art. No. W11421, doi:10.1029/2005WR 004229.

Molotch, N. P., and R. C. Bales. 2006. SNOTEL representativeness in the Rio Grande headwaters on the basis of physiographics and remotely sensed snow cover persistence. Hydrological Processes 20(4):723-739.

Molotch, N. P., M. T. Colee, R. C. Bales, and J. Dozier. 2005. Estimating the spatial distribution of snow water equivalent in an Alpine Basin using binary regression tree models: The impact of digital elevation data and independent variable selection. Hydrological Processes 19(7):1459-1479, doi:10.1002/ hyp.5586.

Mulholland, P. J., J. L.Tank, D. M. Sanzone, W. M. Wolheim, B. J. Peterson, J. R. Webster, and J. L. Meyer. 2000. Food web relationships in a forested stream determined by natural 15N abundance and experimental 15N measurements. Journal of the North American Benthological Society 19:145-157.

National Aeronautics and Space Administration (NASA). 2007. A NASA Earth Science Implementation Plan for Energy and Water Cycle Research: Predict ing Energy and Water Cycle Consequences of Earth System Variability and Change. Draft dated January 12, 2007. Prepared by the Energy- and Water-Cycle Study (NEWS) Science Integration Team. Available on-line at http://www.nasa-news.org/implementation_plan/.

National Research Council (NRC). 1994. Ground Water Recharge Using Waters of Impaired Quality. Washington, D.C.: National Academy Press.

National Research Council (NRC). 1995. Wetlands: Characteristics and Boundaries. Washington, D.C.: National Academy Press.

National Research Council (NRC). 2000a. Ensuring the Climate Record from the NPP and NPOESS Meteorological Satellites. Washington, D.C.: National Academy Press.

National Research Council (NRC). 2000b. From Research to Operations in Weather Satellites and Numerical Weather Prediction: Crossing the Valley of Death. Washington, D.C.: National Academy Press.

National Research Council (NRC). 2002. Estimating Water Use in the United States: A New Paradigm for the National Water-Use Information Program. Washington, D.C.: National Academy Press.

National Research Council (NRC). 2003. Cumulative Environmental Effects of Oil and Gas Activities on Alaska's North Slope. Washington, D.C.: The National Academies Press.

National Research Council (NRC). 2004. Facilitating Interdisciplinary Research. Washington, D.C.: National Academies Press.

National Research Council (NRC). 2005. Flash Flood Forecasting Over Complex Terrain: With an Assessment of the Sulphur Mountain NEXRAD in Southern California. Washington, D.C.: The National Academies Press.

National Research Council (NRC). 2006a. Progress toward Restoring the Everglades: The First Biennial review. Washington, D.C.: The National Academies Press.

National Research Council (NRC). 2006b. Toward an Integrated Arctic Observing Network. Washington, D.C.: The National Academies Press.

National Research Council (NRC). 2006c. CLEANER and NSF's Environmental Observatories. Washington, D.C.: National Academies Press.

National Research Council (NRC). 2007. Earth Science and Applications from Space: National Imperatives for the Next Decade and Beyond. Washington, D.C.: National Academies Press.

National Science Foundation. 2005. Sensors for Environmental Observatories: Report of the NSF-Sponsored Workshop December 2004. Arlington, VA: National Science Foundation. Available on-line at http://www.wtec.org/ seo.

Nelson, F. E., O. A. Anisimov, and N. I. Shiklomanov. 2002. Climate change and hazard zonation in the circum-arctic permafrost regions. Natural Hazards 26(3):203-225.

Njoku, E., and D. Entekhabi. 1996. Passive microwave remote sensing of soil moisture. J. Hydrology 184:101-129.

Njoku, E. G., T. J. Jackson, V. Lakshmi, T. K. Chan, S. V. Nghiem. 2003. Soil moisture retrieval from AMSR-E. IEEE Transactions on Geoscience and Remote Sensing 41(2):215- 229, doi:10.1109/TGRS.2002.808243.

Ntelekos, A. A., K. P. Georgakakos, and W. F. Krajewski. 2006. On the uncertainties of flash flood guidance (FFG): Toward probabilistic forecasting of flash floods. Journal of Hydrometeorology 7(5):896-915.

Nugent, M. A., S. L. Brantley, C. G. Pantano, and P. A. Maurice. 1998. The influence of natural mineral coatings on feldspar weathering. Nature 395:588-591.

Oechel, W. C., G. L. Vourlitis, S. J. Hastings, R. C. Zulueta, L. Hinzman, and D. Kane. 2000. Acclimation of ecosystem CO_2 exchange in the Alaskan Arctic in response to decadal climate warming. Nature 406:978-981.

Oldak, A., T. J. Jackson, P. Starks, and R. Elliott. 2003. Mapping near-surface soil moisture on regional scale using ERS-2 SAR data. International Journal of Remote Sensing 24(22):4579-4598.

Olson, D., N. . Junker, and B. Korty. 1995. Evaluation of 33 years of quantitative precipitation forecasting at the NMC. Weather Forecasting 10:498-511.

Olsson, P. Q., M. Sturm, C. H. Racine, and V. Romanovsky. 2003. Five stages of the Alaskan Arctic cold season with ecosystem implications. Arctic Antarctic and Alpine Research 35(1):74-81.

Paerl, H. W., L. M. Valdes, J. E. Adolf, B. M. Peierls, and L. W. Harding Jr. 2006a. Anthropogenic and climatic influences on the eutrophication of large estuarine ecosystems. Limnology and Oceanography 51: 448-462.

Paerl, H. W., L. M. Valdes, A. R. Joyner, B. L. Peierls, C. P. Buzzelli, M. F. Piehler, S. R. Riggs, R. R. Christian, J. S. Ramus, E. J. Clesceri, L. A. Eby, L. W. Crowder, and R. A. Luettich. 2006b. Ecological response to hurricane events in the Pamlico Sound System, North Carolina and implications for assessment and management in a regime of increased frequency. Estuaries and Coasts 29(6):1033-1045.

Pathak, C. S. (ed.). 2008. Appendix 2-1: Hydrologic Monitoring Network of South Florida Water Management District. G. Redfield, ed. In 2008 Draft South Florida Environmental Report. West Palm Beach, Florida: South Florida Water Management District.

Persidis, A. 2000. Malaria. Nature 18:111-112.

Peter, G., F. J. Kopping, and K. A. Berstis. 1995. Observing and modeling gravity changes caused by soil moisture and groundwater table variations with superconducting gravimeters in Richmond, Florida, U.S.A. Cahiers du Centre Europeen de Geodynamique et de Seismologie 11:147-159.

Peterson, B. J., W. M. Wollheim, P. J. Mulholland, J. R. Webster, J. L. Meyer, J. L. Tank, E. Marti, W. B. Bowden, H. M. Valett, A. E. Hershey, W. H. McDowell, W. K. Dodds, S. K. Hamilton, S. Gregory, and D. D. Morrall. 2001. Control of nitrogen export from watersheds by headwater streams. Science 292(5514):86-90.

Pham, H., E. J. Kim, and A. W. England. 2005. An analytical calibration approach for microwave polarimetric radiometers. IEEE Trans. Geosci. Remote Sens. 43(11):2443-2451.

Phillips, F. M. 1994. Environmental tracers for water movement in desert soils of the American Southwest. Soil Sci. Soc. Am. Journal 58:14-24.

Porter, J., P. Arzberger, H-W. Braun, P. Bryant, S. Gage, T. Hansen, P. Hanson, C-C. Lin, F-P. Lin, T. Kratz, W. Michener, S. Shapiro, and T. Williams. 2005. Wireless sensor networks for ecology. BioScience 55:561-572.

Porter, K. G., and Y. S. Feig. 1980. The use of DAPI for identifying and counting aquatic microflora. Limnol. Oceanogr. 25:943-948.

Potter, K. W. 2006. Small-scale, spatially distributed water management practices: Implications for research in the hydrologic sciences. Water Resources Research 42:W03S08, doi:10.1029/2005WR004295.

Qi, S. L., A. Konduris, D. W. Litke, and J. Dupree. 2002. Classification of irrigated land using satellite imagery, the High Plains aquifer, nominal data 1992. U.S. Geological Survey Water Res. Inv. Report 02-4236. 31 pp.

Ramillien G., F. Frappart, A. Cazenave, and A. Guentner. 2005. Time variations of land water storage from an inversion of 2 years of GRACE geoids. Earth and Planetary Science Letters 235: 283-301.

Reckhow, K., L. Band, C. Duffy, J. Famiglietti, D. Genereux, J. Helly, R. Hooper, W. Drajewski, D. McKnight, F. Ogden, B. Scanlon, and L. Shabman. 2004. Designing Hydrologic Observatories: A Paper Prototype of the Neuse Watershed. A report to the Consortium of Universities for the Advancement of Hydrologic Sciences, Inc. CUAHSI Technical Report Number 6. Washington, D.C.

Reichle, R. H., D. B. McLaughlin, and D. Entekhabi. 2001a. Variational data assimilation of microwave radiobrightness observations for land surface hydrology applications. IEEE Transactions on Geoscience and Remote Sensing 39:1708-1718.

Reichle, R. H., D. Entekhabi, and D. B. McLaughlin. 2001b. Downscaling of radiobrightness measurements for soil moisture estimation: A four-dimensional variational data assimilation approach. Water Resources Research 37(9):2353-2364.

Rignot, E., and R. H. Thomas. 2002. Mass balance of polar ice sheets. Science 297(5586):1502-1506.

Riu, J., A. Maroto, and F. X. Rius. 2006. Nanosensors in environmental analysis. Talanta 69:288-301.

Rockstrom, J., and M. Falkenmark. 2000. Semiarid crop production from a hydrological perspective: Gap between potential and actual yields. Critical Reviews in Plant Sciences 19:319-346.

Rodell, M., and J. S. Famiglietti. 2002. The potential for satellite-based monitoring of groundwater storage changes using GRACE: The High Plains aquifer, Central United States. J. of Hydrology 263:245-256.

Rodell, M., J. S. Famiglietti, J. Chen, S. I. Seneviratne, P. Viterbo, S. Holl, and C. R. Wilson. 2004. Basin scale estimates of evapotranspiration using GRACE and other observations. Geophysical Research Letters 31:L20504, doi:10.1029/2004GL020873.

Rodell, M., J. Chen, H. Kato, J. Famiglietti, J. Nigro and C. Wilson. 2006. Estimating ground water storage changes in the Mississippi River basin using GRACE. Hydrogeology Journal 15(1): 159-166.

Rogers, D. J., and S. E. Randolph. 1991. Mortality rates and population density of tsetse flies correlated with satellite imagery. Nature 351:739-741.

Root, R. A., K. M. Campbell, J. G. Hering, and P. A. O'Day. 2005. Mechanism of Arsenic Sequestration in High-Iron Sediments. American Geophysical Union, Fall Meeting 2005, Abstract B31A-0957.

Ruberg, S., S. Brandt, R. Muzzi, N. Hawley, T. Bridgeman, G. Leshkevich, J. Lane, and T. Miller. 2007. A wireless real-time coastal observation network. EOS Transactions American Geophysical Union 88(28):285-286.

Ruggaber, T. P., J. W. Talley, and L.A. Montestruque. 2007. Using embedded sensor networks to monitor, control, and reduce CSO events: A pilot study. Environmental embedded sensor networks. Environmental Engineering Science 24 (2):172-182.

Sams III, J. I., and G. A. Veloski. 2003. Evaluation of airborne thermal infrared for locating mine drainage sites in the Lower Kettle Creek and Cooks Run basins, Pennsylvania, USA. Mine Water and the Environment 22:85-93.

Scanlon, B. R., and R. S. Goldsmith. 1997. Field study of spatial variability in unsaturated flow beneath and adjacent to playas. Water Resources Research 33:2239-2252.

Scanlon, B. R., R. C. Reedy, D. A. Stonestrom, D. E. Prudic, and K. F. Dennehy. 2005. Impact of land use and land cover change on groundwater recharge and quality in the southwestern USA. Global Change Biology 11:1577-1593.

Scanlon, B. R., I. D. Jolly, M. Sophocleous, and L. Zhang. 2007. Global impacts of conversions from natural to agricultural ecosystems on water resources: Quantity versus quality. Water Resources Research 43, W03437, doi:10.1029/2006WR005486.

Schmugge, T., and P. O'Neill. 1986. Passive microwave soil moisture research. IEEE Trans. Geosci. Remote Sensing GE-24:12-22.

Schmugge, T., and T. Jackson. 1992. A dielectric model of the vegetation effects on the microwave emission from soils. IEEE Trans. Geosci. Remote Sensing 30:757-760.

Schoenung, S. M., and S .S. Wegener. 1999. Meteorological and remote sensing applications of high altitude unmanned aerial vehicles. International Airborne Remote Sensing Conference and Exhibition, 4th/21st Canadian Symposium on Remote Sensing, Ottawa, Canada, June 21-24, pp I-429-I-436.

Seders, L. A., C. A. Shea, M. D. Lemmon, P. A. Maurice, and J. W. Talley. 2007. LakeNet: An integrated sensor network for environmental sensing in lakes. Environmental embedded sensor networks. Environmental Engineering Science 24(2):183-191.

Selker, J. S., L. Thévenaz, H. Huwald, A. Mallet, W. Luxemburg, N. van de Giesen, M. Stejskal, J. Zeman, M. Westhoff, and M. B. Parlange. 2006. Distributed fiberoptic temperature sensing for hydrologic systems. Water Resources Research 42: W12202, doi:10.1029/2006WR005326.

Selley, R. C. 2000. Applied Sedimentology. San Diego, Ca.: Academic Press. 523 pp.

Seo, D. J., V. Koren, and N. Cajina. 2003. Real-time variational assimilation of hydrologic and hydrometeorological data into operational hydrologic forecasting. J. Hydrometeorology 4(3):627-641.

Shi, J., K. Chen, J. J. Van Zyl, K. Kim, and E. G. Njoku. 2002. Estimate relative soil moisture change with multi-temporal L-band radar measurements. Pp. 147-649 in IEEE International Geoscience and Remote Sensing Symposium Volume 1.

Shiklomanov, I. A. 2000. Appraisal and assessment of world water resources. Water International 25:11-32.

Smith, L. C., Y. Sheng, G. M. MacDonald, and L. D. Hinzman. 2005. Disappearing Arctic lakes. Science 308:1429, doi:10.1126/science.1108142.

Snow, R.W., E. McCabe, D. Mbogo, C. Molyneaux, E. Some, V. Mung'ala, and C. Nevill. 1999. The effect of delivery mechanisms on the uptake of bednet reimpregnation in Kifili District, Kenya. Health Policy and Planning 14: 18-25.

Sokol, Z. 2003. MOS-based precipitation forecasts for river basins. Weather and Forecasting 18:769-781.

Sperfslage, J. A., K. P. Georgakakos, T. M. Carpenter, E. Shamir, N. E. Graham, R. Alfaro, and L. Soriano. 2004. Central America Flash Flood Guidance (CAFFG) User's Guide. HRC Limited Distribution Report No. 21. San Diego, Ca.: Hydrologic Research Center. 82 pp.

Spruill, T. B., A. J. Tesoriero, H .E. Mew Jr., K. M. Farrell, S. L. Harden, A. B. Colosimo, and S. R. Kramer. 2004. Geochemistry and characteristics of nitrogen transport at a confined animal feeding operation in a coastal plain agricultural watershed, and implications for nutrient loading in the Neuse River Basin, North Carolina, 1999-2002. Scientific Investigations Report 2004-5283. U.S. Department of the Interior, Washington, D.C.: US Geological Survey.

Steere, D. C., A. Baptista, D. McNamee, C. Pu, and J. Walpole. 2000. Research challenges in environmental observation and forecasting systems. Pp. 292-299 in Proceedings of the 6th Annual International Conference on Mobile Computing and Networking. Boston: Association for Computing Machinery.

Stewart, I. T., D. R. Cayan, and M. D. Dettinger. 2005. Changes toward earlier streamflow timing across western North America. Journal of Climate 18(8):1136-1155.

Strassberg, G., B. R. Scanlon, and M. Rodell. 2007. Comparison of seasonal terrestrial water storage variations from GRACE with groundwater-level measurements from the High Plains Aquifer (USA). Geophysical Research Letters 34, L14402, doi:10.1029/2007GL030139.

Sukhatme, G. S., A. Dhariwal, B. Zhang, C. Oberg, B. Stauffer, and D. A. Caron. 2007. The design and development of a wireless robotic networked aquatic microbial observing system. Environmental Engineering Science 24(2): 205-215.

Sweeney, T. L. 1992. Modernized Areal Flash Flood Guidance. NOAA Technical Report NWS HYDRO 44. Silver Spring, Md.: Hydrology Laboratory, National Weather Service. 21 pp.

Swenson, S., and J. Wahr. 2006. A method for estimating large-scale precipitation minus evapotranspiration from GRACE satellite gravity mission, to appear. Journal Hydrometeor. 7(2):252-270.

Swenson, S. C., P. J.-F. Yeh, J. Wahr, and J. S. Famiglietti. 2006. A comparison of terrestrial water storage variations from GRACE with in situ measurements from Illinois. Geophysical Research Letters 33:L16401, doi:10.1029/2006 GL026962.

Syed, T. H., J. S. Famiglietti, J. Chen, M. Rodell, S. I. Seneviratne, P. Viterbo, and C. R. Wilson. 2005. Total basin discharge for the Amazon and Mississippi River basins from GRACE and a land-atmosphere water balance. Geophysical Research Letter 32:L24404, doi:10.1029/ 2005GL024851.

Syed, T. H., J. S. Famiglietti, M. Rodell, J. Chen, and C. R. Wilson. 2008. Terrestrial hydroclimatology from GRACE and GLDAS. Water Resources Research 44, doi.10.1029/2006WR005779.

Szewczyk, R., E. Osterweil, J. Polastre, M. Hamilton, A. Mainwaring, and D. Estrin. 2004. Habitat monitoring with sensor networks. Communications of the ACM 47(6):34-40.

Tague, C. L., and L. E. Band. 2004. RHESSys: Regional hydro-ecologic simulation system—An object-oriented approach to spatially distributed modeling of carbon, water, and nutrient cycling. Earth Interactions 8:1-42.

Tapley, B. D., S. Bettadpur, J. C. Ries, P. F. Thompson, and M. M. Watkins. 2004. GRACE measurements of mass variability in the Earth system. Science 305:503-505.

Tesoriero, A. J., T. B. Spruill, H. E. Mew Jr., K. M. Farrell, and S. L. Harden. 2005. Nitrogen transport and transformation in a coastal plain watershed: Influence of geomorphology on flow paths and residence times. Water Resources Research 41:W02008, doi:10.1029/2003WR002953.

Trenberth, K. E., L. Smith, T. Qian, A. Dai and J. Fasullo. 2007. Estimates of the global water budget and its annual cycle using observational and model data. J. Hydrometeor. 8(4): 758-769.

U.S. Army Corps of Engineers (USACE). 2002. Forecast-based advance release at Folsom Dam, effectiveness and risks, Phase I. HEC Report PR-48. Davis, Cal.: Army Corps of Engineers, Hydrologic Engineering Center. 93 pp.

U.S. Department of the Interior. 2005. Water 2025: Preventing crises and conflict in the West. Washington, D.C.

U.S. Global Change Research Program (USGCRP). 2003. U.S. National Assessment of the Potential Consequences of Climate Variability and Change Educational Resources Regional Paper: Alaska. Available on-line at http://www.usgcrp.gov/usgcrp/nacc/education/alaska.

Velicogna, I., and J. Wahr. 2006a. Measurements of time-variable gravity show mass loss in Antarctica. Science 311(5768):1754-1756.

Velicogna, I., and J. Wahr. 2006b. Acceleration of Greenland ice mass loss in spring 2004. Nature 443: 329-331, doi:10.1038/nature05168.

Vernon, F., T. Hansen, K. Lindquist, B. Ludaescher, J. Orcutt, and A. Rajasekar. 2003. ROADNET: A real-time data aware system for earth, oceanographic, and environmental applications. EOS Transactions (American Geophysical Union fall meeting supplement) 84: U21A-06.

Vorenhout, M., H. G. van der Geest, D. van Marum, K. Wattel, and H. J. P. Eijsackers. 2004. Automated and continuous redox potential measurements in soil. J. Environ. Qual. 33:1562-1567.

Vörösmarty, C. J., L. Hinzman, B. J. Peterson, D. L. Bromwich, L. Hamilton, J. Morison, V. Romanovsky, M. Sturm, and R. Webb. 2001. The Hydrologic Cycle and its Role in Arctic and Global Environmental Change: A Rationale and Strategy for Synthesis Study. Fairbanks, Alaska: Arctic Research Consortium of the United States (ARCUS). 84 pp.

Wang, J. R., J. C. Shiue, T. J. Schmugge, and E. T. Engman. 1989. Mapping surface soil moisture with l-band radiometric measurements. Remote Sensing of the Environment RSEEA7 27(3):305-312.

Warren, J. K. 2006. Evaporites: Sediments, Resources and Hydrocarbons. Berlin: Springer.

Watson, K. M., Y. Bock, and D. T. Sandwell. 2002. Satellite interferometric observations of displacements associated with seasonal groundwater in the Los Angeles basin. Journal of Geophysical Research 107(B4):2074, doi:10.1029/2001JB000470.

Western, A.W., R. B. Grayson, G. Blöschl, G. Willgoose, and T. A. McMahon. 1999. Observed spatial organisation of soil moisture and relation to terrain indices. Water Resources Research 35(3):797-810.

Western, A. W., S. L. Zhou, R. B. Grayson, T. A. McMahon, G. Blöschl, and D. J. Wilson. 2004. Spatial correlation of soil moisture in small catchments and its relationship to dominant spatial hydrological processes. Journal of Hydrology 286(1-4):113-134.

Wigmosta, M. S., L. Vail, and D. P. Lettenmaier. 1994. A distributed hydrology-vegetation model for complex terrain. Water Resources Research 30: 1665-1679.

Wilker, H., M. Drusch, G. Seuffert, and C. Simmer. 2006. Effects of the near-surface soil moisture profile on the assimilation of L-band microwave brightness temperature. Journal of Hydrometeorology 7(3):433-442.

Wilson, J. L., and H. Guan. 2004. Mountain-block hydrology and mountain front recharge. Pp. 113-117 in Groundwater Recharge in a Desert Environment: the Southwestern United States. J. F. Hogan, F. M. Phillips, and B. R. Scanlon, eds. Water Science and applications Series, Vol. 9. Washington, D.C.: American Geophysical Union.

Wilson, W. J., S. H. Yueh, S. J. Dinardo, S. L. Chazanoff, A. Kitiyakara, F. K. Li, Y. Rahmat-Samil. 2001. Passive active L-and S-band (PALS) microwave sensor for ocean salinity and soil moisture measurements, Geoscience and Remote Sensing. IEEE Transactions 39:1039-1048.

Wood, E. F., and A. Szollosi-Nagy, eds. 1980. Recent Developments in Real-time Forecasting/Control of Water Resources Systems. Pergamon, Oxford. 330 pp.

Wood, W. W., and W. E. Sanford. 1995. Chemical and isotopic methods for quantifying ground-water recharge in a regional, semiarid environment. Ground Water 33:458-468.

Wood, W. W., K. A. Rainwater, and D. B. Thompson. 1997. Quantifying macro-pore recharge: Examples from a semi-arid area. Ground Water 35: 1097-1106.

Woodhouse, B., and T. Hansen. 2003. Meeting the challenges of real-time data transport and integration: HPWREN and ROADNET. Southwest Hydrology 2:16-17.

World Health Organization and UNICEF. 2006. Meeting the MDG drinking water and sanitation target: The urban and rural challenge of the decade. Geneva, Switzerland: WHO. Available on-line at http://www.who.int/water_sanitation_health/monitoring/jmpfinal.pdf.

Wu, W., and S. S. Y. Wang. 2006. Formulas for sediment porosity and settling velocity. J. Hydr. Engrg. 132(8):858-862.

Yang, B. Z., B. Aksak, Q. Lin, and M. Sitti. 2006. Compliant and low-cost humidity nanosensors using nanoporous polymer membranes. Sensors and Actuators B: Chemical 114(1):254-262.

Yao, H., and A. P. Georgakakos. 2001. Assessment of Folsom Lake response to historical and potential future climate scenarios: 2. Reservoir Management. Journal. of Hydrology 249:176-196. Also USGS Open File Report 00-336.

Yao, K., D. Estrin, and Y. H. Hu. 2003. Special issue on sensor networks. European Journal on Applied Signal Processing 2003(4): 319-320.

Yeh, P. J.-F., S. C. Swenson, J. S. Famiglietti, and M. Rodell. 2006. Remote sensing of groundwater storage variations using GRACE. Water Resources Research 42:W12203, doi:10.1029/2006WR005374.

Zehe, E., and G. Blöschl. 2004. Predictability of hydrologic response at the plot and catchment scales: Role of initial conditions. Water Resources Research 40:W10202, doi: 10.1029/2003WR002869.

Zhang, D., L. Zuqin, C. Li, T. Tany, X. Liu, S. Han, B. Lei, and C. Zhou. 2004. Detection of NO_2 down to ppb levels using individual and multiple In_2O_3 nanowire devices. Nano Letters 4(10):1919-1924.

Zhou, Y. H., D. McLaughlin, and D. Entekhabi. 2006. Assessing the performance of the ensemble Kalman filter for land surface data assimilation. Monthly Weather Review 134(8):2128-2142.

Appendix A

Key Water Science Research Questions and Challenges

(Derived from National Research Council reports and meetings)

WATER QUALITY AND QUANTITY

The challenge is to develop an improved understanding of and ability to predict changes in freshwater resources and the environment caused by floods, droughts, sedimentation, and contamination. Important research areas include improving understanding of hydrologic responses to precipitation, surface water generation and transport, environmental stresses on aquatic ecosystems, the relationships between landscape changes and sediment fluxes, and subsurface transport, as well as mapping groundwater recharge and discharge vulnerability.
Grand Challenges in the Environmental Sciences (NRC, 2001)

Validate the water cycle components of climate models. The science questions contained in the water cycle science plan that are related to understanding and predicting variability require an improved understanding of hydrologic processes and their representation in climate models. Therefore, it seems that advances in this area are also fundamental to the water cycle science plan, and the research community is poised to make these advances. Advanced climate change impact assessments are dependent on progress in this area. The path forward in this area requires the identification of the weakest elements in the characterization of the water cycle, and it requires the identification of quantitative improvement goals.
Review of USGCRP Plan for a New Science Initiative on the Global Water Cycle (NRC, 2002)

Scaling of Dynamic Behavior: In varied guises throughout hydrologic science we encounter questions concerning the quantitative relationship between the same process occurring at disparate spatial or temporal scales. Most frequently

perhaps, these are problems of complex aggregation that are confounding our attempts to quantify predictions of large-scale hydrologic processes. The physics of a nonlinear process is well known under idealized, one-dimensional laboratory conditions, and we wish to quantify the process under the three-dimensional heterogeneity of natural systems, which are orders of magnitude larger in scale. Solving these problems will require well-conceived field data collection programs in concert with analysis directed toward "renormalization" of the underlying dynamics. Success will bring to hydrologic science the power of generalization, with its dividends of insight and economy of effort.
Opportunities in the Hydrologic Sciences (NRC, 1991)

Innovative Engineering Approaches for Improving Water Quantity and Quality Management: The research should aim to improve our capabilities in hydrologic forecasting for water resource managers to evaluate and make decisions. Networks of sensors, robotic water quality monitoring sites, realtime data collection, and communication links can be developed into an intelligent environmental control system that will enhance the protection of urban ecosystems and the health and safety of its inhabitants. Such a system can be used as an early warning system and to identify emerging problems such as flooding, lack of water, riparian habitat degradation, and the presence of toxic compounds.
CLEANER and NSF's Environmental Observatories (NRC, 2006)

Land Surface-Atmosphere Interactions: Understanding the reciprocal influences between land surface processes and weather and climate is more than an interesting basic research question; it has become especially urgent because of accelerating human-induced changes in land surface characteristics in the United States and globally. The issues are important from the mesoscale upward to continental scales. Our knowledge of the time and space distributions of rainfall, soil moisture, ground water recharge, and evapotranspiration are remarkably inadequate, in part because historical data bases are point measurements from which we have attempted extrapolation to large-scale fields. Our knowledge of their variability, and of the sensitivity of local and regional climates to alterations in land surface properties, is especially poor. The opportunity now exists for great progress on these issues.
Opportunities in the Hydrologic Sciences (NRC, 1991)

Find solutions to existing and emerging problems involving contaminants in the environment that affect ecosystems and human health. Some environmental problems that affect water resources are of such a magnitude that they are of

national concern and require engineering research based on data collected through observatories. Two such problems are the containment or removal of contaminated sediments and the effects on aquatic and human health of residuals from pharmaceuticals and other household products.
CLEANER and NSF's Environmental Observatories (NRC, 2006)

Sediment Transport and Geomorphology: Erosion, transport, and deposition of sediments in fluvial systems control the very life cycle of rivers and are vulnerable to changes in climate and human landscape alternations. Yet, compared with water quality and quantity information, there is relatively little available information on sediment behavior in river systems, particularly large-order reaches. By advancing basic research on sediment transport detection, quantification of bedload, suspended load, and washload, and monitoring flow velocity and water temperature associated with such sediment transport conditions, the USGS could better detect morphologically significant flows, develop methods to mitigate future problems arising from sediment movement, and play a guiding role in multiagency efforts to deal with the increasingly important national sediment challenges.
River Science at the U.S. Geological Survey (NRC, 2007)

Coordinated Global-scale Observation of Water Reservoirs and the Fluxes of Water and Energy: Regional and continental-scale water resources forecasts and many issues of global change depend for their resolution on a detailed understanding of the state and variability of the global water balance. Our current knowledge is spotty in its areal coverage; highly uneven in its quality; limited in character to the quantities of primary historical interest (namely precipitation, streamflow, and surface water reservoirs); and largely unavailable still as homogeneous, coordinated, global data sets.
Opportunities in the Hydrologic Sciences (NRC, 1991)

Learn how changes in climate, land cover, and land use affect water quantity and quality regimes and their impact on ecosystem health and other uses of water such as for drinking, irrigation, industry, and recreation. Using long-term data, comparative studies, modeling, and experiments, observatory systems can determine pathways of movement of water and solutes through human-dominated landscapes and forecast responses to changes.
CLEANER and NSF's Environmental Observatories (NRC, 2006)

HYDROECOLOGY AND BIOGEOCHEMISTRY

Land Use and Habitat Alteration: Deforestation, suburbanization, road construction, agriculture, and other human land-use activities cause changes in ecosystems. Those changes modify water, energy and material balances and the ability of the biotic community to respond to and recover from stress and disturbance. Actions in one location, such as farming practices in the upper Midwest, can affect areas 1,000 or more miles away because areas are joined by water and nutrient flow in rivers and by atmospheric transport of agrochemicals.
NEON: Addressing the Nation's Environmental Challenges (NRC, 2003)

Chemical and Biological Components of the Hydrologic Cycle: In combination with components of the hydrologic cycle, aqueous geochemistry is the key to understanding many of the pathways of water through soil and rock, for revealing historical states having value in climate research, and for reconstructing the erosional history of continents. Together with the physics of flow in geologic media, aquatic chemistry and microbiology will reveal solute transformations, biogeochemical functioning, and the mechanisms for both contamination and purification of soils and water. Water is the basis for much ecosystem structure, and many ecosystems are active participants in the hydrologic cycle. Understanding these interactions between ecosystems and the hydrologic cycle is essential to interpreting, forecasting, and even ameliorating global climate change.
Opportunities in the Hydrologic Sciences (NRC, 1991)

Ecological Implications of Climate Change: Human-induced climate warming and variability strongly affect individual species, community structure and ecosystem functioning. Changes in vegetation in turn affect climate through their role in partitioning radiation and precipitation at the land surface. Climate-driven biological impacts are often only discernable at a regional-continental scale. Regional changes in ecosystem processes affect global water and carbon cycles. Therefore, a national approach to understanding biological response to climate variability and change is required.
NEON: Addressing the Nation's Environmental Challenges (NRC, 2003)

[Grand Challenges include:]

Biogeochemical Cycles: The challenge is to further our understanding of the Earth's major biogeochemical cycles, evaluate how they are being perturbed by human activities, and determine how they might better be stabilized. Important research areas include quantifying the sources and sinks of the nutrient elements and gaining a better understanding of the biological, chemical, and physical factors regulating transformations among them; improving understanding of the interactions among the various biogeochemical cycles; assessing anthropogenic perturbations of biogeochemical cycles and their impacts on ecosystem functioning, atmospheric chemistry, and human activities, and developing a scientific basis for societal decisions about managing these cycles; and exploring technical and institutional approaches to managing anthropogenic perturbations.

Invasive species: Invasive species affect virtually every ecosystem in the United States, and can cause substantial economic and biological damage. The identification of potentially harmful invasive species, the early detection of new species as invasion begins, and the knowledge base needed to prevent their spread require a comprehensive monitoring and experimental network and a mechanistic understanding of the interplay of invader, ecosystem traits and other factors including climate and land use that determine invasiveness.

Grand Challenges in the Environmental Sciences (NRC, 2001)

The nation is spending billions on riverine restoration and rehabilitation projects, yet the science underlying these projects is not currently well understood and thus the approaches and their effectiveness vary widely. Therefore, a fundamental challenge is to quantitatively understand how rivers respond physically and biologically to human alterations from dredging to damming, and to specifically address: What are the required "environmental flows" (i.e., flow levels, timing, and variability) necessary to maintain a healthy river ecosystem? And which biota and ecological processes are most important and/or sensitive to changes in river systems?

River Science at the U.S. Geological Survey (NRC, 2007)

How can local riverine ecosystem processes be scaled from habitat patches across river reaches to produce basin-wide predictive capabilities? (I.e., how can we estimate regional aquatic ecosystem processes over river basins?)

COHS workshop on "Towards Integration of Hydrologic and Ecological Sciences," West Palm Beach, Florida, October 2000.

HEALTH

Algal Blooms and Water-Borne Infectious Diseases: The rapid proliferation of toxic or nuisance algae, termed harmful algal blooms (HAB), can occur in marine, estuarine and freshwaters, and are one of the most scientifically complex

and economically significant water issues facing the United States today.
Earth Science and Applications from Space: National Imperatives for the Next Decade and Beyond (NRC, 2007)

Vector Borne and Zoonotic (VBZ) Disease: VBZ diseases, such as malaria, dengue, and filariasis are believed responsible for millions of deaths and tens of millions of illnesses annually. The introduction and spread of West Nile virus through North America by mosquitoes during the past five years and recent concerns about the world-wide dissemination of H5N1 avian influenza are key recent examples where large human populations have come at risk over extensive geographic regions in short periods of time by these VBZ diseases. Attempts to control VBZ disease epidemics with limited available resources are hindered by the ability to prioritize and target areas for intervention. The major goal of such [remote sensing] efforts is to establish relationships between environmental conditions, as monitored by satellites, and risk to human populations from VBZ diseases. This goal requires improved characterization of the earth's land use, ecological changes and changing weather, at finer spatial and temporal scales.
Earth Science and Applications from Space: National Imperatives for the Next Decade and Beyond (NRC, 2007)

Infectious Disease and the Environment: The challenge is to understand ecological and evolutionary aspects of infectious diseases; develop an understanding of the interactions among pathogens, hosts/receptors, and the environment; and thus make it possible to prevent changes in the infectivity and virulence of organisms that threaten plant, animal, and human health at the population level. Important research areas include examining the effects of environmental changes as selection agents on pathogen virulence and host resistance; exploring the impacts of environmental change on disease etiology, vectors, and toxic organisms; developing new approaches to surveillance and monitoring; and improving theoretical models of host-pathogen ecology.
Grand Challenges in the Environmental Sciences (NRC, 2001)

Appendix B

Planning, Designing, Operating, and Utilizing the Results from an Integrated Observational-Modeling System

The planning, design, operation, and utilization of an integrated observational-modeling system involves many elements, or stages. Some of these are scientific or technological, whereas others are organizational or social. Eight such elements are summarized here.

These include (1) defining goals, which may include specific "deliverables" for a narrowly defined research project or flexible targets when the project is established for broader and potentially changing uses; (2) building an initial team with appropriate expertise to define and oversee accomplishment of the goals, but often allowing the team to change over time as a project evolves; (3) designing the project to achieve the goals, either specifically or with flexibility to allow for multiple-use data; (4) collecting and validating the data, integrating and validating new data collection methods as appropriate over time; (5) organizing the large data sets for a variety of different uses; (6) integrating observations across sensors and networks; (7) merging the integrated observations with models and model validation; and (8) delivering the information products from the integrated observations and merged observation-modeling to those applying them to flood and drought forecasting, water management planning, disaster response, source water protection, and other areas.

These eight elements often must be addressed in an iterative fashion as a project evolves. For example, once information is delivered, managers or other members of the scientific community may suggest changes to an ongoing project to meet additional or changing needs. The elements are discussed briefly below and are either explicitly or implicitly part of the studies summarized in Chapter 4. As the case studies were initiated for different reasons and are ongoing, each shines a light on different elements or sets of elements.

(1) Defining project goals and "deliverables" is, of course, part of any pro-

posal process. In some cases, these goals appear to be deceptively straightforward. For example, for the Comprehensive Everglades Restoration Plan (see Chapter 4), "The overarching objective of the Plan is the restoration, preservation, and protection of the South Florida ecosystem while providing for other water-related needs of the region, including water supply and flood protection" (Water Resources Development Act of 2000). However, enormous amounts of time and energy have been—and continue to be—invested to define what "restoration" constitutes, and what the end points might be. Major multidisciplinary scientific initiatives wrestle with integrating multiple objectives, such as understanding fundamental processes such as streamflow generation, investigating scaling relationships of observations over time and space, understanding behavior under extreme conditions, and developing new instrumentation. The more multidisciplinary the project the more difficult—and more critical—it is to establish one's goals at the onset. In many cases, it is essential to allow the goals to change over time as new methods are developed, new ideas evolve, and new researchers add to both the needs and the capabilities of the project.

(2) Building a strong, interdisciplinary team, when the project spans disciplines, is as essential as it is supremely challenging. As one participant in an NRC workshop expressed it, "A 'multidisciplinary' team is put together and they work in isolation until the very end, when they fight." Some of the difficulties are neither scientific, nor institutional, but personal. Keys to success in putting together an interdisciplinary group include finding colleagues who work at institutions that have policies and practices that lower barriers to interdisciplinary scholarship, and are willing to "immerse themselves in the languages, cultures, and knowledge of their collaborators" (NRC, 2004).

(3) Designing a project to achieve the goals set out, whether narrow or broad, specific or flexible, is the next step. An overall approach for the particular needs of interdisciplinary collaborations is described in Benda et al. (2002) as follows:

> [T]he success of interdisciplinary collaborations among scientists can be increased by adopting a formal methodology that considers the structure of knowledge in cooperating disciplines. For our purposes, the structure of knowledge comprises five categories of information: (1) disciplinary history and attendant forms of available scientific knowledge; (2) spatial and temporal scales at which that knowledge applies; (3) precision (i.e., qualitative versus quantitative nature of understanding across different scales); (4) accuracy of predictions; and (5) availability of data to construct, calibrate, and test predictive models. By definition, therefore, evaluating a structure of knowledge reveals limitations in scientific understanding, such as what knowledge is lacking or what temporal or spatial scale mismatches exist among disciplines.

This process, if followed at the project formulation stage, can be used to construct "solvable problems", and involves building consensus among team members with respect to precision requirements, scales of analysis, and disciplinary expertise needed. A further advantage is that it leads to a feedback loop to examine whether the project goals and "deliverables" as originally conceived may need to be modified or rejected (Benda et al., 2002).

(4) Collecting and validating data as it relates to sensors is covered in Chapter 2. Other aspects of data collection and validation, while central to integrated observing systems, is beyond the scope of this report. This is clearly an enormous field by itself. The Environmental Protection Agency's *Field Operations Manual for Wadeable Streams* (EPA, 1998) details protocols for a wide variety of activities from stream discharge measurements to periphyton sampling. The U.S. Geological Survey lists method, sampling, and analytical protocols for a variety of dissolved solutes and suspended materials at http://water.usgs.gov/nawqa/protocols/ methodprotocols.html, and biological sampling, habitat, and laboratory protocols at http://water.usgs.gov/nawqa/protocols/bioprotocols.html.

(5) The organization of data-sets is another important step. It is dealt with in this report in the context of smart sensors and sensor networks, which can help to avoid the collection of large amounts of time series data at times when little change is occurring in the measured parameter, and to reduce data where this is deemed useful. With satellites, the amount of data generated can be enormous—on the order of a terabyte per day for a single satellite (NRC, 2000a) and thus overall several thousand terabytes per year. In fact, in many cases the size of the data archives is growing faster than we can derive information from them; NASA's Earth science data holdings increased by a factor of six between 1994 and 1999 and then doubled again from 1999 to 2000 (Climate Change Science Program, 2003). However, this is also a highly parameter-specific activity, and it is difficult to generalize principles without a specific context.

(6) Integrating observations across sensors and networks: Currently vast amounts of environmental and water-related information are collected daily by a wide range of sensors, and these data are being used widely for water management, water-quality monitoring, flood hazard forecasting, and so forth. Examples of such applications are provided in the case studies in Chapter 4. Sensors range from snow measurements taken manually by the National Weather Service to experimental embedded network sensors to control storm water discharge. And in between, there is a tremendous range of operational and experimental sensor platforms or stations that collect, store, and transmit data in a variety of ways. Currently, most sensor platforms are unable to communicate directly with each other, and there is a lack of inter-operability among data networks, for the most part, which will be discussed below.

New developments in sensor technology are occurring rapidly, both in the ability to obtain measurements in new novel ways (for example, through biosensors for water quality) and in transmitting the information through long-lived self-configuring embedded networked sensing systems. In brief, these networks embed a computational intelligence in the environment, linking sensor pods through wireless technology in a manner that allows the network to conduct adaptive monitoring and real-time control. Development of new sensors from nanosensors to new satellite-based systems was also described in Chapter 2.

(7) Merging observations with models: Data sets are frequently assimilated into models, both to provide model-based forecasts (e.g., upper air observations used for weather forecasting; precipitation and river stage observations to forecast flood stages) and to predict variables not well measured (e.g., nonpoint pollution runoff, terrestrial evaporation).

(8) Using results from an integrated observational-modeling system: Data and model products have no value unless they are used. They can only be used if they can be easily discovered, acquired, and understood in a timely manner to those who wish to apply them to practical issues such as flood forecasting, water availability modeling, and ecological flows, as inputs to decisionmaking. The communication and delivery of data and information to such end users is the back-bone to a beneficial integrated system. New "web-based hydrologic services" are being developed at the University of Texas by Professor David Maidment under National Science Foundation funding, and similar applications with remote sensing at the University of Illinois by Professor Praveen Kumar (Box 3-1). These nascent activities facilitate data discovery, acquisition, and integration need to be further developed and integrated with users through demonstration projects; and other competing approaches need to be developed and evaluated through similar means.

Appendix C

A Complementary National Research Council Study on Earth Science and Applications from Space

As noted in the Preface, a separate National Research Council study titled "Earth Science and Applications from Space" and commonly known as the "decadal survey" (NRC, 2007; http://www.nap.edu/catalog/11820.html), was conceived at about the same time that the present study was funded and begun. The decadal survey was much broader and encompassed the totality of the geosciences, including not only water resources and the global hydrologic cycle but also land-use change, ecosystem dynamics, biodiversity, weather, climate, human health, and the solid Earth sciences. The decadal survey recommended that "the nation should execute. . . an integrated in situ and space-based observing system" and noted specifically that "[t]he scientific challenge posed by the need to observe the global water cycle is to integrate in situ and space-borne observations to quantify the key water cycle state variables and fluxes." In practice, however, it focused on space-based observations with little attention to the details of in-situ measurements. At its heart was a list of 17 recommended satellite missions to support national needs for research and monitoring of the Earth system during the decade 2005-2015.

The four highest ranked water cycle-related missions of the decadal survey, all of which were incorporated into the 17 recommended missions, were

1. The approved, but delayed, Global Precipitation Measurement (GPM) Mission, which would provide diurnal estimates of precipitation at a spatial resolution sufficient to resolve major spatial variations over land and sea;
2. A soil moisture mission that would provide estimates of soil moisture over most of the Earth;
3. A surface-water mission that would provide observations of the variability of water stored in lakes, reservoirs, wetlands, and river channels, and would support estimates of river discharge; and

4. A cold season mission that would estimate the water storage of snow packs, especially in spatially heterogeneous mountainous regions that are the source areas for many of the world's most important rivers.

Taken together, these four missions would form the basis for a coordinated effort to observe most components of the surface-water cycle globally.

In addition to these four, other missions recommended in NRC (2007) primarily for other areas of geoscience, but with direct application to water science and applications, included missions that would estimate water vapor transport, sea ice and glacier mass balance, groundwater and ocean mass, and inland and coastal water quality. Details of these missions are given in NRC (2007).

These recommended missions would clearly contribute greatly to the kinds of scientific investigations illustrated by the case studies in Chapter 4 and proposed for hydrologic, environmental engineering, and ecological observatories. For example, the improved estimates of light rain and snow would benefit virtually all of the case studies and proposed observatories. Since soil moisture (and its freeze/thaw state) is the key variable that links the water, energy, and biogeochemical cycles, and is a key determinant of evapotranspiration (NRC, 2007), the same could be said for the soil moisture mission. The cold season mission would be directly applicable to snowy regions of highly variable topography, such as that illustrated in "Mountain Hydrology in the Western United States." The surface-water mission, based on a radar altimeter, would be able to capture the spatial dynamics of many periodically flooded areas, thus assisting in studies of water-related disease, for example, "Water and Malaria in Sub-Saharan Africa." It might also help capture the slight variations in water elevations that drive the ecology and nutrient cycling described in "Monitoring the Hydrology of the Everglades in South Florida."

The other four missions cited in the water cycle chapter of the decadal report would also have obvious impacts on studies that attempt to integrate observations at different scales. For example, the groundwater and ocean mass mission, known as the Gravity Recovery and Climate Experiment (GRACE) follow-on mission, could lead toward better evapotranspiration estimates by improving the terrestrial water storage change, as suggested in the Southern High Plains case study. And the inland and coastal water quality mission would have a steerable 250 m resolution spectrometer designed to quantify the response of marine ecosystems to short-term physical events—one of the primary topics of the Neuse River Basin and Estuary Study. Details of these missions are given in NRC (2007).

The water-related satellite missions recommended by the decadal study are consistent with the vision, findings, and recommendations of this study. The measurements and retrievals from these missions should contribute to the vision offered in this report of ground-to-space integrated observations systems feeding into decision-support systems, if they are strategically combined with a multi-agency, ground-based measurement strategy.

Appendix D

Biographical Sketches Committee on Integrated Observations for Hydrologic and Related Sciences

KENNETH W. POTTER, *Chair*, is a professor of civil and environmental engineering at the University of Wisconsin, Madison. His teaching and research interests are in hydrology and water resources, including hydrologic modeling, estimation of hydrologic risk, estimation of hydrologic budgets, watershed monitoring and assessment, and hydrologic restoration. Dr. Potter is a past member of the Water Science and Technology Board and has served on many of its committees. He is a Fellow of the American Association for the Advancement of Science and the American Geophysical Union. He received his B.S. in geology from Louisiana State University and his Ph.D. in geography and environmental engineering from Johns Hopkins University.

ERIC F. WOOD, *Vice Chair*, is a professor in the Department of Civil Engineering and Operations Research, Water Resources Program, at Princeton University. His areas of interest include hydroclimatology with an emphasis on land-atmosphere interaction, hydrologic impact of climate change, stochastic hydrology, hydrologic forecasting, and rainfall-runoff modeling. Dr. Wood is an associate editor for *Reviews of Geophysics*, *Applied Mathematics and Computation: Modeling the Environment*, and *Journal of Forecasting*. He is a member of the Climate Research Committee and the Panel on Climate Change Feedbacks. He is a former member of the Water Science and Technology Board and Board on Atmospheric Science and Climate's Global Energy and Water Cycle Experiment panel. Dr. Wood received an Sc.D. in civil engineering from Massachusetts Institute of Technology in 1974.

ROGER C. BALES joined the University of California (UC), Merced, as professor of engineering in June 2003, and is one of UC Merced's inaugural faculty.

Dr. Bales received his B.S. from Purdue University in 1974, an M.S. from the UC, Berkeley in 1975, and his Ph.D. from the California Institute of Technology in 1985. He worked as a consulting engineer from 1975 to 1980, and was professor of hydrology and water resources at the University of Arizona from 1984-2003. He has published extensively in diverse fields of research including snow hydrology, alpine hydrology and biogeochemistry, polar snow and ice, contaminant hydrology, and water quality. At the University of Arizona Dr. Bales served as director of the NASA-supported Regional Earth Science Applications Center (RESAC), deputy director of the NSF-supported Science and Technology Center on Sustainability of Semi-Arid Hydrology and Riparian Areas (SAHRA), and principal investigator on the NOAA-supported Climate Assessment for the Southwest Project (CLIMAS). He is actively involved in research in the southwestern United States, Greenland, and Antarctica. Dr. Bales is a Fellow of the American Association for the Advancement of Science (AAAS), of the American Geophysical Union (AGU), and the American Meteorological Society (AMS). He serves on a number of advisory committees and professional society boards.

LAWRENCE E. BAND is chair and Voit Gilmore Distinguished Professor of Geography at the University of North Carolina. He received a B.A. in geography from the State University of New York at Buffalo, summa cum laude, in 1977, and an M.S. and Ph.D. in geography from the University of California, Los Angeles, the latter in 1983. His research interests are watershed hydrology, ecology, geomorphology, geographic information systems (GIS), remote sensing, structure, function, and dynamics of watershed systems. Dr. Band combines field measurement and observation of hydrological and ecological variables with development and application of distributed simulation models, GIS, and remote sensing techniques. His projects are particularly concerned with the integration and coupling between water, carbon, and nutrient cycling and transport with watersheds, and the interactions of human behavior as part of watershed ecosystems. He is chair of the Consortium of Universities for the Advancement of Hydrologic Science Committee on Hydrologic Observatories.

ELFATIH A. B. ELTAHIR is a professor in the Department of Civil and Environmental Engineering at the Massachusetts Institute of Technology (MIT). Dr. Eltahir received the Presidential Early Career Award for Science and Engineering. He was nominated by the National Aeronautics and Space Administration for the honor based on his work in hydroclimatology. The award cites Professor Eltahir's "outstanding accomplishment in hydrology and hydroclimatology by combining theory and remote sensing observations to better understand the links between the biosphere and the atmosphere and their implications for regional water resources in the tropics." He is a past member of the executive committee of the hydrology section of the American Geophysical Union (AGU). He was also editor of *Geophysical Research Letters* from 1998 to 2001. Dr. Eltahir received a B.S. in 1985 from University of Khartoum, an M.S. in 1988 from National University of Ireland, and an S.M. and an Sc.D. from MIT, both in 1993.

ANTHONY W. ENGLAND is associate dean of the College of Engineering and professor of Electrical Engineering and Computer Science at the University of Michigan. Dr. England's research interests are development and field calibration of Land Surface Process (LSP) models of land-atmosphere energy and moisture fluxes for northern prairie and Arctic tundra, assimilation of satellite microwave brightness data for improved estimates of moisture profiles in prairie and Arctic soils and snow, and development of microwave radiometers for use on towers, aircraft, and satellites. He is a geophysicist by training. Dr. England received a B.S. in earth sciences from MIT in 1965, an M.S. in geology and geophysics from MIT in 1965, and a Ph.D. in geophysics from MIT in 1970.

JAMES S. FAMIGLIETTI is an associate professor of Earth System Science at the University of California, Irvine. He holds a B.S. in geology from Tufts University, an M.S. in hydrology from the University of Arizona, and an M.A. and a Ph.D. in civil engineering from Princeton University. Major areas of current service are as an editor for *Geophysical Research Letters*, as co-chair of the AGU Hydrology Section Remote Sensing committee, and as the Graduate Advisor for Earth System Science. His research concerns the role of hydrology in the coupled Earth system. Areas of current activity include hydrologic and climate system modeling for studies of land-ocean-atmosphere interaction, satellite remote sensing of soil moisture and terrestrial water storage, soil moisture variability and scaling, and global change impacts on water resources and hydrology-vegetation interaction.

KONSTANTINE P. GEORGAKAKOS is the managing director of the Hydrologic Research Center in San Diego, California. He is also an adjunct full professor with the Scripps Institution of Oceanography of the University of California, San Diego, and an adjunct full professor with the Department of Civil and Environmental Engineering of the University of Iowa. Previously, he was an associate professor at the University of Iowa and with the Iowa Institute of Hydraulic Research, as well as a research hydrologist with the National Weather Service. He holds M.S. and Sc.D. degrees from the Massachusetts Institute of Technology. Honors and awards include the Presidential Young Investigator Award from the National Science Foundation and the NRC-NOAA Associateship Award from the National Research Council. He has authored or co-authored more than 100 refereed publications in the areas of hydrology, hydrometeorology, and hydroclimatology. As part of the science cooperation and technology transfer activities of the Hydrologic Research Center, he led the design and implementation of an operational multispectral satellite rainfall estimation system for the Nile River Basin, the operational hydrologic forecast system for Peru, a prototype integrated climate-hydrology-reservoir forecast-management system for the water resources of northern California, and the regional operational flash flood warning system for the seven countries of Central America. He is a consultant for the United Nations Food and Agriculture Organization, and he has been associate editor of the American Society of Civil Engineers' *Journal of Engineering Hydrology*, the *Journal of Hydrology* and *Advances in Water Resources*. He serves as the U.S. Expert in Hydrologic Modeling for the World Meteorological Organization Commission for Hydrology (Working Group on Applications).

DINA L. LOPEZ is an associate professor of geology at Ohio University. Dr. Lopez's research interests include acid-mine drainage and its impact on the receiving waters; geochemistry and hydrogeology of hydrothermal systems and lakes affected by volcanic and anthropogenic inputs in Central America; diffuse soil degassing in volcanic areas, including the flux of CO_2 and radon; and modeling of the coupled fluid flow and heat transfer in geological systems. She has received numerous international fellowships, and multiple outstanding teaching awards at Ohio University. Dr. Lopez received a B.Sc in chemistry from the University of El Salvador in 1975, a M.Sc. in physics from Virginia Tech in 1979, and a Ph.D. in geology from Louisiana State University in 1992. Before working at Ohio University, she held a postdoctoral fellowship in hydrogeology at the University of British Columbia.

DANIEL P. LOUCKS works in the application of systems analysis, economic theory, ecology and environmental engineering to problems in regional development and environmental quality management including air, land, and water resource systems. He is a member of the National Academy of Engineers. At Cornell University, he served as chair of the Department of Civil and Environmental Engineering from 1974 to 1980, and as associate dean for research and graduate studies in the College of Engineering from 1980 to 1981. Since 1969 he has served as a consultant to private and government agencies and various organizations of the United Nations, the World Bank, and North Atlanta Treaty Organization on regional water resources development planning throughout the world. He has served on various committees of the NRC, including the International Institute for Applied Systems Analysis liaison committee and, most recently, the Committee on Restoration of the Greater Everglades Ecosystem. The Secretary of the Army appointed him to the U.S. Army Corps of Engineers Environmental Advisory Board in 1994. He served as vice chair and chair from 1995 to 1998, and received the Commander's Award for Public Service in 1998. Dr. Loucks received his M.F. in forestry from Yale University and his Ph.D. in environmental engineering from Cornell University.

PATRICIA A. MAURICE is a professor of civil engineering and geological sciences at the University of Notre Dame and is director of the university's interdisciplinary Center for Environmental Science and Technology. Dr. Maurice's-research focuses on microbial, trace metal, and organic interactions with mineral surfaces from the atomic scale up to the scale of entire watersheds. Her research encompasses the hydrology and biogeochemistry of freshwater wetlands and mineral-water interactions, the remediation of metal contamination, and global climate change. Dr. Maurice received her Ph.D. in aqueous and surface geochemistry from Stanford University.

LEAL A. K. MERTES was a professor of geography at the University of California, Santa Barbara. She was an interdisciplinary scientist with B.S. degrees in both geology and biology from Stanford University and M.S. and Ph.D. degrees in geology from the University of Washington. Her research investigated the geomorphic and hydrologic processes responsible for the development of wetlands and floodplains in large river systems and across watersheds and to develop remote sensing techniques for analysis of wetlands and water properties. Dr. Mertes' research also included the study of large rivers and the use of digital data for analysis of ecosystem dynamics. She taught courses on digital image processing of remote sensing data and rivers. For the National Academies she served on the Committee to the International Contributions for Scientific, Educational and Cultural Activities (ICSECA) Program, the Committee on International Organizations and Programs, and the International Advisory Board. Dr. Mertes held memberships in the Geological Society of America and the American Geophysical Union. She received her Ph.D. in geological sciences from the University of Washington. She died on September 30, 2005.

WILLIAM K. MICHENER is associate director of the Long Term Ecological Research Network Office in the Department of Biology at the University of New Mexico. His activities include network coordination, informatics research, and international training. He served as program director of ecology in the Division of Biological Sciences at the National Science Foundation (NSF) from 1999-2000. At NSF, he served as senior biological sciences program officer in the Biocomplexity Working Group, and participated in numerous cross-disciplinary and interagency initiatives. His research interests include the ecology of natural and anthropogenic disturbances, and ecological informatics. He has published three books dealing with ecological informatics and more than 70 journal articles and book chapters. He is co-director of the Project Office and chair of the Facilities & Infrastructure Committee for the proposed National Ecological Observatory Network (NEON).

BRIDGET R. SCANLON is a senior research scientist in the Bureau of Economic Geology and also teaches courses in the geology and civil engineering departments at the University of Texas at Austin. Her expertise lies in unsaturated zone hydrology, soil physics, environmental tracers, and numerical simulations to quantify subsurface flow in arid regions. She served on the NRC Committee on Ward Valley. She has served as a consultant to the Nuclear Waste

Technical Review Board and associate editor for *Hydrogeology Journal*. She regularly gives short courses on methods of estimating rates of groundwater recharge. Dr. Scanlon received her Ph.D. in geology at the University of Kentucky.